DEVELOPING EMISSION BASELINES FOR MARKET-BASED MECHANISMS: A CASE STUDY APPROACH

July 2001

U.S. Department of Energy
National Energy Technology Center
626 Cochrans Mill Road
Pittsburgh, PA 15236

PREFACE

This report was sponsored by the U.S. Department of Energy's National Energy Technology Center (NETL), as part of NETL's continuing efforts to explore and address the issue of global climate change. It deals with the international global climate change agreement reached in Kyoto, Japan in December 1997. More specifically, it addresses the issue of developing the emission baseline under market-based mechanisms aimed at achieving the emission reduction goals to arise under future international climate change agreements.

This report was prepared by Chris Minnucci, Jette Findsen, Chris Mahoney, Sarah Billups, and Michael Mondshine of Science Applications International Corporation (SAIC), under the guidance of James Ekmann (NETL).

TABLE OF CONTENTS

EXECUTIVE SUMMARY i
    Chapter 1. Introduction i
    Chapter 2. Approaches for Quantifying the Emission Baseline iv
    Chapter 3. Recommended Generic Procedure for Establishing Emission Baselines vi
    Chapter 4. Emissions Baseline Development for the Indian Power Plant Efficiency
        Improvement Project vii
    Chapter 5. Emissions Baseline Development for an Integrated Gasification Combined
        Cycle (IGCC) Power Project in China ix
    Chapter 6. Fuel Cells in Argentina xi
    Chapter 7. Critique of the Project Analysis xii
    Chapter 8. Summary and Conclusions xiii

1. INTRODUCTION 1
    1.1 Background: Market Mechanisms 1
        1.1.1 Implications of the Market Mechanisms 3
    1.2 Objectives and Report Organization 5

2. APPROACHES FOR QUANTIFYING THE EMISSION BASELINE 9
    2.1 Introduction 9
    2.2 Defining Baselines 9
        2.2.1 Criteria for Baseline Development for Market-Based Emission Reduction
            Activities 10
            2.2.1.1 Additionality 11
            2.2.1.2 Level of Error 12
                2.2.1.2.1 Additionality Classification Errors 12
                2.2.1.2.2 The Treatment of Time 16
            2.2.1.3 Transparency 17
            2.2.1.4 Transaction Costs 18
            2.2.1.5 Summary 19
    2.3 Review of Baseline Options 19
    2.4 The Project-Specific Approach 20
        2.4.1 Additionality and the Project-Specific Approach 21
            2.4.1.1 Economic Feasibility Approach 23
                2.4.1.1.1 Considerations in the Application of the Economic
                    Feasibility Approach 23
            2.4.1.2 Non-Economic Barriers 26
                2.4.1.2.1 The Knowledge Barrier 26
                2.4.1.2.2 Lack of Access to Financing 27
                2.4.1.2.3 Considerations in the Application of the Non-Economic
                    Barriers Test 28

2.4.2 Options for Computing Emission Baselines 30
    2.4.2.1 Historical Baselines 30
    2.4.2.2 Modified Baselines 31
    2.4.2.3 Dynamic Baselines for Evaluating Long-lived Projects 32
2.4.3 Evaluation of the Project-Specific Approach 33
2.5 The Benchmarking Approach 33
  2.5.1 Level of Benchmark Aggregation 35
    2.5.1.1 National Level Benchmarks 35
    2.5.1.2 Sector Level Benchmarks 36
    2.5.1.3 Sub-sector Level Benchmarks 38
    2.5.1.4 Global/Regional Benchmarks 40
  2.5.2 Options for Computing Benchmarks 40
    2.5.2.1 Historical Benchmarks 41
    2.5.2.2 Projected Benchmarks 41
    2.5.2.3 Normative Benchmarks 41
  2.5.3 Additionality and the Benchmark Approach 42
  2.5.4 Evaluation of the Benchmark Approach 42
2.6 The Modified Technology Matrix Approach 43
  2.6.1 The Early Version of the Technology Matrix 44
  2.6.2 A Modified Version of the Technology Matrix 44
  2.6.3 Additionality and the Modified Technology Matrix 45
    2.6.3.1 The Economic Feasibility Test 46
    2.6.3.2 The Market Penetration Test 46
  2.6.4 Quantifying Stipulated Baselines 48
    2.6.4.1 Dynamic Versus Static Baselines 50
  2.6.5 Evaluation of the Modified Technology Matrix Approach 51
2.7 Conclusion 52

3. GENERIC PROCEDURE FOR ESTABLISHING EMISSION BASELINES 53
3.1 Assessment of the Three Proposed Procedures 53
  3.1.1 Potential Error Sources in The Benchmarking Approach 53
  3.1.2 Applicability of the Project-Specific and Technology Matrix Approaches
    56
3.2 Generic Baseline Development Guidelines 58
  3.2.1 Approach Selection Criteria 59
  3.2.2 Development of the Technology Matrix 61
  3.2.3 Some General Comments Concerning Standardization of the Project-Specific
    Approach 63
3.3 Summary 64

4. EMISSIONS BASELINE DEVELOPMENT FOR THE
   INDIAN POWER PLANT EFFICIENCY IMPROVEMENT PROJECT 65

4.1 Introduction 65
4.2 Project Description 65
   4.2.1 The GEP Project 65
   4.2.2 The Hypothetical Indian Power Plant Efficiency Improvement Project . . . 68
4.3 Emission Baseline Development 69
   4.3.1 Evaluation of Baseline Options 69
   4.3.2 Additionality 70
      4.3.2.1 The Financial Barrier 70
      4.3.2.2 The Knowledge Barrier 74
      4.3.2.3 Temporal Considerations 76
   4.3.3 Baseline Development 77
      4.3.3.1 Establishing the Qualitative Baseline 77
      4.3.3.2 Quantifying the Baseline 83
         4.3.3.2.1 Estimating the "Split" Between the Generation Increase
            and the Fuel Consumption Reduction 85
         4.3.3.2.2 The Treatment of Time in the Equations 86
         4.3.3.2.3 Simplifications Incorporated in the Algorithms 87
         4.3.3.2.4 Miscellany 89
4.4 Summary 90

5. EMISSIONS BASELINE DEVELOPMENT FOR AN INTEGRATED GASIFICATION
   COMBINED CYCLE (IGCC) POWER PROJECT IN CHINA 91
5.1 Project Description 91
   5.1.1 Background 91
   5.1.2 The Project 92
5.2 Emission Baseline Development 94
   5.2.1 Evaluation of Baseline Options 94
   5.2.2 Additionality 95
      5.2.2.1 Economic Feasibility 96
      5.2.2.2 Market Penetration 99
      5.2.2.3 Temporal Considerations and Additionality 100
5.3 Establishing the Benchmark 101
   5.3.1 Qualitative Considerations 102
   5.3.2 Variable Versus Constant Benchmark 105
   5.3.3 Quantifying the Benchmark 107
   5.3.4 Regional Considerations 109
   5.3.5 Temporal Considerations 109
5.4 Summary 111

6. FUEL CELLS IN ARGENTINA 114
6.1 Project Description 114

6.1.1 Background 114
6.1.2 The Project 114
6.2 Emission Baseline Development 116
6.2.1 Evaluation of Baseline Options 116
6.2.2 Additionality 117
6.2.2.1 Economic Feasibility of SOFC. 117
6.2.2.2 Market Penetration of SOFC in Argentina. 119
6.2.2.3 Temporal Considerations and Additionality. 119
6.3 Establishing the Benchmark 120
6.3.1 The Most Likely Alternative 120
6.3.2 Benchmark Subcategories 121
6.3.3 Benchmark Metric 122
6.3.4 Temporal Considerations 123
6.4 Summary 125

7. CRITIQUE OF THE PROJECT ANALYSIS 126
7.1 The Implications of Supply-Demand Disequilibriums 126
7.1.1 Electricity Supply-Demand Imbalances 127
7.1.1.1 The Indian Power Plant Efficiency Improvement Project 127
7.1.1.2 The Chinese IGCC Project 129
7.1.2 Fuel Supply Demand Imbalances 130
7.2 Weaknesses in the Analysis of the Indian Power Plant Efficiency Improvement Project
. . . . 131
7.2.1 Additionality 131
7.2.1.1 The Financial Barrier Argument 131
7.2.1.2 The Knowledge Barrier Argument 133
7.2.1.3 Some Conclusions Concerning the Additionality of the Indian
Power Plant Efficiency Improvement Project 133
7.2.2 The Emissions Baseline 134
7.3 Qualitative Error Assessment 135
7.3.1 Biases 136
7.3.1.1 Biases Arising from Baseline Estimation 136
7.3.1.2 Biases Arising from Additionality Classification Errors 137
7.4 Towards a Solution 137
7.4.1 Reducing Additionality Classification Errors 138
7.4.2 Accommodating Biases in Emission Baseline Estimates 140
7.4.2.1 Placing International Agreements in Perspective 141

8. CONCLUSIONS AND RECOMMENDATIONS FOR FURTHER WORK 143

    8.1 Report Summary and Conclusions 143

BIBLIOGRAPHY 145

GLOSSARY 147

LIST OF TABLES

Table 2.1. Comparison of Benefits, Uses, Advantages, and Disadvantages of Baseline Approaches 22

Table 2.2. Comparison of Uses, Advantages, and Disadvantages of the Levels of Benchmark Aggregation 38

Table 2.3. Example of a Portion of the Technology Matrix 47

Table 2.4. Estimating Emission Reductions Using the Modified Technology Matrix 48

Table 3.1. Criteria for Selecting an Approach to Baseline Development, for the Electricity Generation Sector 60

Table 5.1. Comparative Economic Factors for U.S.-Based Power Generation Systems 97

Table 5.2. Levelized Costs for IGCC Versus PC in China 97

LIST OF FIGURES

Figure 2.1. Baseline Option Continuum: Tradeoff Between Transparency, Error, and Transaction Costs 9

Figure 2.2. Projects Qualifying as Additional Under a Rigorous Additionality Test 15

Figure 2.3. Projects Qualifying as Additional Under a Relaxed Additionality Test 15

Figure 2.4. Classification Errors and Lost Opportunities Under a "Mid-Range" Additionality Test 15

Figure 2.5. Investor Preferences for Qualifying Non-Additional Projects 16

Figure 2.6. Use of a Rigorous Additionality Test to Minimize Emission Reduction Estimation Error 17

Figure 2.7. Determining Additionality Under the Project Specific Approach 24

Figure 2.8. Different Methods for Establishing the Benchmark 36

Figure 4.1. Determining Additionality of the Indian Power Plant Efficiency Improvement Project 72

Figure 4.2. Establishing a Qualitative Baseline Under the Project-Specific Approach 80

EXECUTIVE SUMMARY

Chapter 1. Introduction

Concern about increasing atmospheric concentrations of carbon dioxide and other greenhouse gases, and the potential impact of these increases on the earth's climate, has grown significantly over the past decade. This concern has led to a series of international meetings and agreements seeking to stabilize atmospheric greenhouse gas concentrations. In 1992, at Rio De Janeiro, the Framework Convention on Climate Change (UNFCCC) was signed by more than 160 countries, including the United States. There was widespread agreement among the signatories on the potential negative effects of climate change under a business-as-usual future. Under the convention, the developed countries (referred to as Annex I countries) were assigned primary responsibility for addressing the climate change issue. However, between 1992 and 1997, Parties to the Convention strongly disagreed over what policy instruments should be used to curb global climate change, and what, if any, targets and timetables should be set for achieving emission reductions.

A break in the negotiations occurred in late 1997. At the Third Conference of Parties[1] held in Kyoto, Japan in December 1997, a series of firm emission reduction targets were agreed to by the Parties The industrialized countries agreed to reduce their greenhouse gas emissions by an average of 5.2 percent from 1990 levels by 2008-2012. The U.S. agreed to limit its emissions to seven percent below 1990 levels. Since then negotiations on implementing these reductions have stalled and the Protocol has not been ratified. However, major progress in allowing the use of market mechanisms to achieve emission reduction goals occurred at Kyoto. Emissions trading and a new concept where entities can acquire credits for emission reduction activities were among the market-based mechanisms under consideration. This report is concerned exclusively with the latter.

One example of a market mechanism is the Clean Development Mechanism (CDM). A CDM activity is defined in Article 12 of the Kyoto Protocol as a project between a developed country and a developing country that provides the developing country with project financing and technology, while assisting the developed country in meeting its emission reduction commitments. Under the CDM, projects yield emission reductions credits. The share of these credits that is accrued by entities in industrialized countries may be applied towards their own emission reduction goals. Credits are verified and authenticated units of greenhouse gas reductions from abatement or sequestration projects. They are issued pursuant to the review and certification of each project, by an operational organization (CDM Board) to be defined by the Conference of Parties. To obtain credits in a market mechanism environment, a project will likely need to meet the following criteria: 1) voluntary participation; 2) real, measurable, and long-term mitigation of climate change; 3) benifits in addition to what would have occurred in the absence of the project activity (additionality); and 4) projects

---

[1]The Conference of the Parties (COP) is the supreme body of the United Nations Framework Convention on Climate Change established in 1992. The body meets annually and its primary responsibility is to oversee the implementation of the Convention. The Sixth Conference of Parties (COP6) is scheduled for November, 2000.

must contribute to the sustainable development goals of the host country. Sustainable development benefits are not limited to, but should include the following: income benefits, employment benefits, benefits to the local environment, quality of technology transferred, and contributions to the local capacity to sustain and build on the projects. As these factors vary among countries, it will be the responsibility of the individual host country government to determine whether projects satisfy national sustainable development objectives.

It is important to recognize that the market mechanisms are *not* designed to reduce global greenhouse gas emissions beyond emission reduction targets such as those specified in the Kyoto Protocol. Rather, the purpose of market mechanisms is to increase flexibility and reduce the *costs* associated with meeting emission reduction targets. The CDM for example, provides for a one-to-one trade between developed and developing countries. Thus, at least in the ideal, market-based projects will yield no net change in global emissions. In short, it is the emission reduction targets specified in some future international emission reduction agreements, and not the market mechanisms, that will act as the driving force for reducing global greenhouse gas emissions.

While from the perspective of Annex I countries the purpose of the market mechanisms is to reduce the costs associated with reducing emissions, they may be perceived differently by the developing countries (non-Annex 1 countries). The latter may view the market mechanisms as a means of fostering their sustainable development goals. Under the market mechanisms, developing countries will have access to favorable financing for certain types of projects; market-based projects will be required to not only reduce emissions, but also meet the sustainable development goals of the host countries. Although various formal criteria have been suggested for assessing a project relative to the goal of sustainable development, in the final analysis each host country will judge for itself whether or not proposed projects meet its sustainable development goals.

If they can be implemented as originally envisioned, the market mechanism will provide a "win-win" opportunity: it will enable developing countries to further their sustainable development goals while at the same time reducing the developed countries' emission reduction costs. But while the market mechanisms in general may, at least in theory, wed these two different goals in a mutually beneficial union, in practice tension may arise between the two objectives. In particular, the developed countries have voiced their concern that they may become a kind of "dumping ground" for high-risk, uneconomic experimental technologies that developed countries are unwilling to develop domestically. Yet it is precisely advanced, high-risk, marginally economic technologies that most clearly qualify as additional under market mechanisms. Projects utilizing conventional technologies, which may better meet the host countries' goals, may also in many cases fail to meet the test for additionality. This underlying tension, and more importantly how it is addressed, may to a large extent decide the success or failure of the market mechanisms. Success will require the striking of a delicate balance between developed and developing country goals.

This report represents a step towards the development of protocols for the estimation of greenhouse gas emission reductions resulting from potential market mechanism projects undertaken in the power sector. It deals specifically with the difficult and complex problem of developing emission baselines for carbon offset projects. Although exchanging credits for emission reduction activities and technologies is a relatively new concept, much has already been written about it. The literature has identified and developed a number of approaches to emission baseline estimation under the market mechanism concept, and the pros and cons of each approach have been assessed and reviewed at some length (see bibliography). However, the literature has to a large extent considered baseline estimation only in the abstract. Different estimation approaches have been compared and contrasted, but, to date, few attempts have been made to *apply* these approaches.

The primary goal of this report is to help advance the discussion of baseline estimation procedures by *applying* alternative estimation approaches to three hypothetical project case studies. Thus, following an analysis of the three major baseline methodologies under consideration for the market mechanisms, we apply two of these methodologies to hypothetical emission reduction projects.

One of these projects, designed to improve the efficiency of coal-fired power plants in India, is based on an actual ongoing project that is being sponsored by the U.S. Agency for International Development, along with the U.S. Department of Energy's National Energy Technology Laboratory (NETL), the Tennessee Valley

> **Key Definitions**
>
> Project-specific Approach: involves the tailoring of a separate baseline estimation methodology to each individual project, based on a detailed analysis of the project's defining characteristics.
>
> Modified Technology Matrix Approach: a set of technologies is pre-qualified as additional based on a consideration of their economics and current market penetration; baseline emission rates are stipulated.
>
> Benchmark Approach: a set of stipulated baseline emission rates are provided for different countries, sectors, or sub-sectors.

Authority, the Electric Power Research Institute, and India's National Thermal Power Corporation (NTPC). In this report, an emissions baseline for the project is developed using the project-specific approach. The other two projects - an Integrated Gasification-Combined Cycle (IGCC) project in China and a fuel cell project in Argentina - are of a more hypothetical nature, although similar projects are being considered by these countries and others. The authors utilize the modified technology matrix approach to evaluate both of these projects.

Our case studies involve two distinct steps. First, for each project, we adopt the viewpoint of the project developers. From this subjective viewpoint, we attempt to develop as persuasive an argument as possible in support of the additionality (the notion that a project would only happen because of market mechanism incentives) of the three projects. In addition, we seek to be as rigorous as possible in our attempts to develop accurate, reliable project baselines and benchmarks.

Then, following the completion of the three project analysis, we adopt an objective viewpoint, and submit the analysis to a critique. By applying the baseline development approaches in as rigorous a manner as possible, and then subjecting the results of these analysis to an objective critique, our goal is to identify the strengths and weaknesses of each approach under a "best-case" scenario. More specifically, our objective is to identify potential error sources for each approach, and to provide at least a rough qualitative characterization of these error sources as to their magnitude and potential for biasing global-level emission reduction estimates. Only by approaching baseline development with some rigor can we ensure that the error assessment excludes errors resulting from a mere lack of diligence, and focuses on those more formidable errors that may be inherent to the nature of the estimation approaches, and to the nature of the baseline development problem itself.

We apply two different baseline development approaches – the project-specific approach and the modified technology matrix approach – because each approach has its applications, and significant benefits may be gained by combining both approaches in a set of flexible protocols, rather than choosing one over the other for all circumstances. Owing to the complexity and cost associated with the project-specific approach, the modified technology matrix approach is more applicable to the China and Argentina projects. The India project, on the other hand, can satisfy the additionality requirement only under the project specific approach because this project does not easily lend itself to standardization.

Chapter 2. Approaches for Quantifying the Emission Baseline

To evaluate the options for developing emission baselines under a market mechanism environment, we first examine the characteristics and criteria for developing greenhouse gas emission baselines under the UNFCCC. Emission baselines represent the standard from which a measure of valid emission reductions or carbon sequestration is established. The baseline can either be derived from a forecast of emissions of the actual activity to be replaced, or on a specific set of emission data collected from relevant sectors within the economy. Once the baseline has been constructed, the emissions associated with the proposed market-based project are calculated and subtracted from the baseline to determine the actual emission reductions of the project.

The development of emission baselines that are accurate and incur low transaction costs is crucial to enhance market mechanism participation and, at the same time, ensure that the credits developed have a positive environmental impact. In the literature on baseline development, four general requirements have been proposed to promote these objectives. The first and most important requirement refers to the issue of additionality; that is, the question whether the activity would occur in the absence of market mechanism incentives. Some amount of emission reduction activities are bound to happen without the implementation of an international GHG emissions treaty and forecasts of these have already been included in the baseline against which potential global reduction targets were determined. As a result, non-additional market-based activities cannot be counted as offset projects without increasing emissions above the global target. Hence, all projects that apply for credit must demonstrate that they are additional to what would otherwise have occurred. The second requirement stems from the need to ensure accuracy in estimating the emission baseline in order to promote credibility. Two issues in particular have an impact on the level of error. These include (1) the treatment of additionality as a criterion for project certification and (2) the consideration of temporal issues in the development and quantification of baselines. As a third requirement, baseline approaches should provide a transparent (i.e., standardized, clearly defined, and easily replicable) methodology for estimating baselines to facilitate increased participation and ensure the credibility of the emission credits. Finally, an effective baseline methodology should minimize transaction costs to encourage the inclusion of a maximum number of market-based projects.

Considering these four criteria, there are trade-offs between the objectives of ensuring accuracy, transparency, and low transaction costs. To increase transparency and reduce transaction costs a certain level of standardization in the application of the baseline approach is required. However, as baselines become more standardized, the level of error in estimating credits increases. This chapter provides an overview of three baseline methodologies, the project-specific approach, the benchmark approach, and the modified technology matrix approach, and discusses the benefits and disadvantages of each approach with respect to how they respond to each of the four requirements outlined above.

The project-specific approach to baseline development is based on an extensive estimate of total GHG emissions with and without the market-based project. Following this approach, additionality is assessed through an evaluation of a project's economic feasibility and an examination of possible non-financial barriers to project implementation and the baseline is quantified using a historic and dynamic analysis. Projects applying the project-specific approach will be dealt with on a case-by-case basis by a certification board and the entities involved in the project. National programs implementing the Activities Implemented Jointly (AIJ) Pilot Phase have relied solely on the project-specific approach for project evaluation. This approach has received extensive criticism for its high transaction costs and the level of complexity involved in baseline development. However, as the project-specific approach makes every effort at determining what would have happened in the absence of market mechanism incentives, it is potentially the most accurate method for setting baselines.

The second methodology is the benchmark approach. This approach relies on an average, median, or other metric derived from a defined aggregate or category (such as a specific region, sector, or technology) to determine the amount of emissions reduced by a given project. Based on the performance of this aggregate, a benchmark is then developed, which projects must improve upon in order to generate valid emission reductions. Thus, the benchmark is based on a comparison of emission rates alone. Benchmarks may be aggregated at a national, sector, sub-sector, or global/regional level, and can be quantified by using a historical, projected, or normative method. Eliminating the use of a site-specific, case-by-case estimation of emissions with and without the project increases transparency and reduces transaction costs. However, the benchmarking approach does not address the issue of additionality separately from the construction of the emission baseline, raising doubts about the environmental integrity of the credits produced.

As an alternative approach to developing standardized emission baselines, we developed the modified technology matrix approach. The matrix consists of a selected country-specific list of greenhouse gas abating technologies that correspond with the sustainable development goals of the host country. Stipulated emission baselines are then determined for each technology on the list. As the sustainable development objectives of individual countries differ depending on the resources and general development objectives of the nation, it will be the responsibility of national host governments to determine which technologies should be considered for inclusion in the matrix. Sustainable development criteria are likely to include such factors as social and economic impacts, quality of technology transferred, environmental benefits, emissions reduction efficiency, and project feasibility.

However, for a technology to be included on the list, it must also be subjected to an additionality test. This test is based on an examination of the commercial viability and market penetration of the technology and will ensure that non-additional technologies are not included in the list of qualifying technologies. Once it has been proven that a technology is in fact additional, a baseline will be developed for that specific technology based on the emissions performance of a select group of comparable technologies within that country. Individual projects applying for emissions credit will then simply demonstrate that the proposed project technology is already listed on the technology matrix, and then use the stipulated baseline from the matrix to calculate the emission reductions of the project. Both the additionality status of the technology and the baseline against which emission credits are compared will be updated regularly. Thus, by introducing a level of standardization to the baseline development process while at the same time subjecting technologies to a rigorous additionality test, the modified technology matrix represents the middle-ground between the objectives of ensuring accuracy and promoting participation in the market mechanisms.

Chapter 3. Recommended Generic Procedure for Establishing Emission Baselines

Two primary conclusions are drawn based on the authors' assessment of the three proposed procedures. First, although it should substantially reduce transaction costs, the benchmarking approach has a significant disadvantage: at best, it addresses the issue of additionality in an indirect and unreliable manner. Second, the choice between the project-specific approach and the modified technology matrix approach is best made on a project-by-project basis, because each approach offers significant advantages over the other depending on the specific circumstances.

The benchmarking approach offers project sponsors significant opportunities for gaining emission reduction credits without reducing emissions. In fact, the benchmark approach strongly favors investment in non-additional projects at the expense of additional projects, for two reasons. First, a numeric benchmark will, at best, prove a crude screen for additionality, which could ultimately lead to the mis-classification of many non-additional projects as additional (and vice versa). Second, because non-additional projects will, by definition, tend to be more economically viable than additional projects, project developers will preferentially invest in the mis-classified non-additional projects at the expense of truly additional projects.

Unlike the benchmark approach, both the project-specific and modified technology matrix approaches are designed to directly address the issue of additionality. In the case of the project-specific approach, the viability of each individual project, without market mechanism incentives, is assessed using such means as economic feasibility analysis and project barrier analysis, while the modified technology matrix approach involves a direct assessment of the commercial viability of individual technologies. Furthermore, both the project-specific approach and the modified technology matrix approach provide reasonably reliable means for estimating the baseline.

However, neither the project-specific approach nor the modified technology matrix approach is a panacea. Exclusive reliance upon one or the other approach may result in significant lost opportunities, as a result of either the expense of implementing the project-specific approach or the automatic disqualification of all projects involving conventional technologies under the modified technology matrix approach. However, a flexible protocol incorporating both approaches will enable the application of the optimal approach in each specific situation.

To ensure appropriate selection between the project-specific and modified technology matrix approaches under a flexible protocol concept, several guidelines need to be developed. The technology matrix should be the default procedure for analyzing all projects involving the installation of new generating capacity utilizing one of the qualifying technologies. There are three exceptions to this rule. First, the approach cannot be used in host countries for which a list of qualifying technologies, or an appropriate set of benchmarks, has not been developed. Second, project developers should be allowed to utilize the project-specific approach to develop their own baseline, if they so desire, and if they can demonstrate, to the satisfaction of a review board, that the baseline thus estimated is more accurate, for their particular project, than the sectoral benchmark. Finally, if the new capacity is being developed primarily to replace existing capacity or generation, rather than to meet new demand, *and* if it is possible to readily identify the existing capacity or generation being replaced, then the emissions from this existing capacity/generation should be used as the baseline rather than a sector benchmark.

For all projects involving non-qualifying technologies or conventional technologies, the project-specific approach must be utilized. However, projects involving the retrofitting of advanced, qualifying technologies may utilize the technology matrix to establish additionality while utilizing the project-specific approach to establish the baseline.

In the following three sections, the three case studies are summarized. Then, our critique of the three project analysis is presented. Finally, the report's main conclusions and recommendations are presented.

Chapter 4. Emissions Baseline Development for the Indian Power Plant Efficiency Improvement Project

The Indian Power Plant Efficiency Improvement project is based on an ongoing project – the Greenhouse Gas Pollution Prevention Project (GEP). The efficiency improvement activities are being conducted under the Efficient Coal Conversion (ECC) component of the GEP project. The primary sponsor of the GEP project is the U.S. Agency for International Development (USAID); in addition, the project team includes the U.S. Department of Energy's National Energy Technology Center (NETL), the Electric Power Research Institute (EPRI), the Tennessee Valley Authority (TVA), and India's National Thermal Power Corporation (NTPC). The goal of the project is to reduce carbon dioxide emissions by improving the efficiency of existing coal-fired power plants.

The power plant energy efficiency improvement project involves systematic performance monitoring and diagnostic testing of the boilers, turbines, condensers, and auxiliary equipment at NTPC's coal-fired power plants. Through industry-standard tests specific plant components are identified that are operating at less than design or optimal efficiency. Corrective actions include capital improvements to worn equipment, procedural changes in plant operations, training of NTPC personnel, and dissemination of knowledge and information gained through the project.

Our hypothetical project is in effect a replication of the activities being performed under the GEP project. Specifically, we assume that these activities are extended to the coal-fired power plants owned by India's various State Electricity Boards. The hypothetical project is referred to in this report as the Indian Power Plant Efficiency Improvement Project, to distinguish it from the real-world GEP project.

The project-specific approach is selected as the baseline development approach for the hypothetical project. The project-specific approach is preferable for two reasons. First, the project involves conventional technology and improvements to an existing power plant. Second, the additionality of the project cannot be demonstrated based solely on a consideration of technology.

Under the project-specific approach, a project's additionality is demonstrated either through an economic feasibility analysis or by providing evidence of non-economic barriers (financial barriers or knowledge barriers) that would prevent the project from being undertaken without the market mechanism incentives. For this project, additionality cannot be demonstrated through an economic feasibility analysis. The project is designed to yield significant efficiency improvements at relatively low costs, and is designed to be replicable throughout the Indian power sector. At first, one may think that the project would not qualify as additional based on the grounds that it would be economic without market mechanism assistance. Nevertheless, certain barriers do exist that would prevent the project from being fully replicated throughout India until it receives the favorable development assistance from the U.S. sponsors.

To identify these barriers and demonstrate additionality, a non-economic barrier analysis is performed. The first step in this analysis is to determine if a financial barrier to project implementation exists. India's State Electricity Boards (SEBs) are in very poor financial health, leading to the tentative conclusion that the SEBs will not fund similar improvements without some form of development assistance.

Other barriers to project implementation may also exist. It is argued that SEB personnel lack the technical knowledge and training necessary to implement the project on their own, demonstrating the existence of a knowledge barrier. Based on this argument, we again conclude that the project qualifies as additional.

After having established that the Indian power plant efficiency improvement project satisfies the additionality criterion, the project's emission baseline is estimated. Under the project-specific approach, the baseline represents a projection of what emissions would have been "but for the project." To answer the question of what would have happened had the Indian efficiency project not been undertaken, the authors follow a standardized step-by-step procedure to identify the most likely alternative emissions scenario to the project. Thus, it is determined that the most likely alternative to the efficiency improvements throughout India would be that utility customers would have relied more heavily on self-generation. In India, electric generating capacity is insufficient to meet demand and many utility customers use diesel generators to backup the grid. Because the efficiency improvements result in an increase in the total power available to end users, as well as a reduction in the fuel consumed for generation, we conclude that in the absence of the project, diesel generators at the point of electricity consumption might be utilized more heavily. Furthermore, we conclude that the project reduces emissions at the affected power plants as well as at the backup generators. A flexible set of algorithms has been specified to quantify the emission reductions at each or both of these two sources.

Chapter 5. Emissions Baseline Development for an Integrated Gasification Combined Cycle (IGCC) Power Project in China

The Integrated Gasification Combined Cycle (IGCC) project in the People's Republic of China (PRC) has been selected because of the interest of the Chinese government in building an IGCC demonstration project and developing domestic capability to produce IGCC technology. The construction of a commercial-scale demonstration IGCC plant by 2000 has been listed as a priority under the PRC's Agenda 21 program. Our hypothetical IGCC market-based project in China will consist of two units, adding 300 and 400 MW of new generation capacity to the grid. The IGCC power plant could be placed in one of the provinces facing severe power shortages, such as Shandong, Gansu, Henan, Wuinghai or Sichuan. The project would rely on imported technology obtained through a combination of direct purchase and technology transfer. The targeted unit efficiency will be 43 percent or higher. If market mechanism incentive become available, project financing would most likely be obtained through a combination of Chinese support and funding from international lending institutions and private investors interested in obtaining emissions credits in exchange for their assistance.

This IGCC project is analyzed using the modified technology matrix approach for two reasons: 1) it involves the construction of new generating capacity and 2) IGCC is an emerging technology that is not yet commercial in China. In Chapter 5, we focus on *qualifying* IGCC technology, and *developing* the stipulated benchmark for inclusion in the technology matrix, rather than *applying* the matrix to the particular project. In this way, we assess IGCC technology in general without reference to the specific project under consideration.

We begin with an assessment of the additionality of IGCC technology. The additionality of IGCC technology may be demonstrated by evaluating the *economic feasibility* and the *market penetration* of the technology. If the technology is found to be unable to compete economically with existing technologies on the market and has failed to reach even a minimal level of market penetration, the technology will qualify as additional. This evaluation should be based on the specific circumstances of the host country and the technology matrix should be developed on a country-by-country basis. Ideally, the technology should qualify only if it meets both the economic feasibility and the market penetration tests. In China, IGCC technology clearly is not commercial, and unless favorable financing is provided through market mechanisms, it will continue to be viewed as too expensive. Moreover, the market penetration rate of the technology in China is zero, and the country has only recently developed the capacity to construct IGCC units domestically. Hence IGCC technology is clearly additional and should be included in China's technology matrix.

Having considered the issue of additionality, the next step is to estimate the emissions baseline. Given a fully developed technology matrix, this is a simple and straightforward process. A stipulated benchmark will be provided for all participating countries and pre-qualifying technologies; the project developers need only identify the appropriate benchmark for their project, and use it as the basis for their emissions baseline. Because a matrix has not yet been established, we focus on the estimation of a benchmark for IGCC projects in China. We begin by addressing the question: what is the most likely alternative to the particular technology under consideration? China needs every megawatt of capacity possible to meet new demand needs arising from the expected expansion of the economy of the next 20 years. It seems prudent to conclude that, in general, new IGCC power plants in China will serve to meet new or currently unmet demand. In light of this conclusion, the next question becomes how would this demand have been met in the absence of this IGCC project? After analyzing several alternatives, we conclude that in general, the counterfactual (the existing project to be replaced) for an IGCC project in China will be a coal-fired power plant utilizing conventional technology, of roughly comparable size to the IGCC plant and located on the same site.

We consider the possibility of using a function, rather than a constant, as the benchmark to capture variations in the counterfactual power plant's heat rate (and hence, its emissions) with factors such as coal quality and utilization. However, owing to limitations in gathering detailed data within this study, it is not clear that the utilization of a functional approach to benchmark determination would represent a significant improvement over the use of a constant. Therefore, the use of a constant rather than a variable benchmark is recommended. Furthermore, a benchmark value based on heat rate (Btus per kilowatt-hour), rather than an emissions rate (pounds of carbon dioxide per kilowatt-hour), is recommended because it will enable us to consider variations in the coal's emission factor across different coal ranks. In other words, separate emission factors (in pounds carbon dioxide per mmBtu) for each rank of coal will be developed and applied to a single benchmark heat rate to determine baseline emissions. The use of rank-specific emission factors is warranted because there are small, but statistically-significant, differences in emission factors for different coal ranks.

Despite our rejection of the functional approach to benchmark development, we recognize that the benchmark should not remain static with respect to time. In other words, the benchmark constant should be updated on a periodic basis to reflect changes or improvements in the operating efficiencies of new coal-fired power plants. It is believed that re-estimation once every five years will be sufficient to keep the benchmark up to date.

Chapter 6. Fuel Cells in Argentina

The proposed Argentinian Fuel Cell project, like the Chinese IGCC project, is hypothetical in nature. At present, there are no operating fuel cell generators in Argentina, nor are there any demonstration projects. However, the Argentinian government has a strong interest in fuel cell technology, primarily for off-grid applications to meet growing rural electricity demand. Much of Argentina's electricity demand growth comes from rural areas that rely heavily on diesel generators to meet their power needs. Our hypothetical project will involve the use of a solid oxide fuel cell (SOFC) as an off-grid generator for a rural village.

As was the case for the China IGCC project, the modified technology matrix approach to emission baseline development is utilized. Clearly, this is a new capacity project utilizing an advanced qualifying technology, which leads us to the modified technology matrix approach. Once again, we focus on the *development* of the technology matrix for fuel cell technology in Argentina rather than its *application* to this particular project. We develop the technology matrix for SOFC technology in Argentina, by (1) demonstrating the additionality of this technology and (2) developing an appropriate emissions benchmark for the technology.

The additionality of SOFC technology in Argentina can be readily demonstrated based on both the economic feasibility and the market penetration of SOFCs. At present, although the application of SOFC technology is growing in the distributed generation market, it is still not considered a proven or mature technology. In fact, the operational performance of SOFC systems is still being tested, although results achieved so far promise acceptable component life characteristics. Moreover, the cost of SOFC technology has still not reached a level where it is competitive on the electricity market.

With regard to market penetration, to date, no fuel cells of any type have been installed in Argentina. Nor have SOFCs achieved significant market penetration levels on a worldwide basis. The technology has been installed at numerous test sites and research facilities. However, these activities have come about mainly through public research support and other incentives. In conclusion, SOFCs have not yet penetrated the Argentine or world markets on a significant scale, providing another indicator that the technology is additional and qualifies for inclusion in the technology matrix.

Having addressed the additionality issue, we now turn to the development of the emissions benchmark for SOFC technology in Argentina. We begin with the following basic question: what is the most likely alternative to the qualifying technology? Recalling Argentina's electricity demand growth in rural areas, the country's interest in utilizing fuel cells to meet this demand growth, and the country's heavy reliance on diesel generators to meet current rural electricity demand, we conclude that the most typical alternative to stationary fuel cell projects is likely to be diesel generators. Therefore, we base the emissions benchmark on the emissions characteristics of diesel generators in Argentina.

In the case of SOFCs in Argentina, we propose that the average emissions rate for new diesel generators (in pounds CO_2 per kilowatt-hour) be used as the benchmark. We have decided on new diesel units for two reasons: 1) any fuel cell project in Argentina will be a new capacity project and 2) Argentina's interest in fuel cells is primarily linked to meeting new demand growth in rural areas, leading to the conclusion that an SOFC project will displace new diesel generators rather than existing generators. We propose that units installed within the past five years be used as the basis for the benchmark. By multiplying the average heat rate of new diesel generators with the diesel fuel emissions factor, the appropriate benchmark emissions rate can be readily derived. The emissions baseline for any particular SOFC project, in a given year, could then be computed by multiplying the benchmark emissions rate by the amount of electricity (in kWh) generated by the fuel cell(s).

Whenever possible, the benchmark should be based on actual heat rate data for operating diesel generators. However, in some instances, such as smaller diesel generators owned by the end users, it may well prove difficult if not impossible to obtain the required heat rate data from the generator owners. In these cases, it may be necessary to use heat rate estimates provided by manufacturers for specific models, in combination with market share data for the different models.

Like the China IGCC project, a constant, as opposed to a functional, benchmark will be utilized with respect to all variables except one – time. To ensure that the benchmarks remain a realistic indicator of current conditions, it will be necessary to update them on a regular basis. A new set of benchmarks, to be applied to new projects only, will be derived every five years. At the same time, the benchmarks to be applied to *ongoing* projects would also be updated every five years. The use of time sensitive benchmarks will capture any changes in the counterfactual emissions rates, thereby reducing the potential for biases in the emission reduction estimates.

Chapter 7. Critique of the Project Analysis

This section summarizes our critique of the three project analysis with a view towards identifying their strengths and weaknesses, and drawing out the lessons that might be learned from this case study. We address a number of specific issues: 1) fundamental project analysis difficulties arising from the characteristics of developing economies, 2) weaknesses in the analysis of the Indian Power Plant Efficiency Improvement project, and 3) the potential for errors, both random and systematic, in the estimation of project baselines.

A number of factors unique to energy markets in the developing world hold implications for baseline development. The most important characteristics are the persistent, chronic supply-demand imbalances. There are two types of imbalances particularly important to the analysis of market-based projects in the power sector: electricity imbalances and fuel imbalances. Electricity imbalances proved to be a major complicating factor in the case of the India project. Because of these imbalances, efficiency improvements may result in increased generation as well as reduced fuel consumption. The potential additional generation is, in turn, expected to displace small end-use diesel generators located at numerous industrial and commercial establishments throughout India. Obtaining reliable emissions-related data on these generators will be a very difficult undertaking.

As for fuel supply-demand imbalances, the basic concern is that if the fuel demands of power plants are going unmet, then market-based projects designed to reduce power plant fuel demand may have *no impact* on fuel consumption or emissions. Rather, such projects may simply reduce the size of the supply-demand gap. In countries facing fuel supply shortages, consumption is determined not by demand but by supply constraints; projects that reduce demand without addressing the supply constraints will have no impact on emissions. For example, if a solar-power irrigation pump is installed on a farm in India, and the farm disconnects from the grid, the electricity is simply used elsewhere by another consumer.

Our critique of the India project uncovered significant weaknesses in the arguments used to support the project's additionality. We use the non-economic barrier approach to establish that the project is additional; however, both the financial barrier argument and the knowledge barrier argument are weak in certain key respects. The basic weakness in the financial barrier argument is that it is not specific to the project at hand. We conclude that the SEBs' weak financial situations would prevent them from taking on the project themselves, but, by this same logic, we would have to conclude that the SEBs are precluded from undertaking all projects. Yet clearly the SEBs do take on some projects. The basic weakness in the knowledge barrier argument is that it relies too heavily on generalities, anecdotal information, and a limited amount of data that might not be available to project developers in actual circumstances.

Other potential error sources were identified through our analysis of the Indian project, and the Chinese and Argentine projects. These include uncertainties surrounding the establishment of the qualitative counterfactual; potential biases inherent in certain critical data; and potential biases arising from additionality classification errors.

To deal with these potential errors, we consider two possible approaches: (1) reducing biases through the development and application of rigorous baseline development procedures and (2) accommodating biases within the framework of future international agreements. The first option may be the best available option for reducing additionality classification errors in a cost-effective manner (i.e., the modified technology matrix with a stringent technology qualification/additionality criteria), but the second option may prove preferable for dealing with the biases arising during the baseline estimation process. This latter option might, for example, involve the addition of constant or variable adjustment factors to the emission targets established under future international accords, to account for systematic errors in market mechanism emission reduction estimates.

Chapter 8. Summary and Conclusions

This report presents a detailed analysis of two alternative approaches to estimating project emission baselines under a market-based mechanism environment: the project-specific approach and the modified technology matrix approach. The project-specific approach was applied to a heat rate improvement project at coal-fired power plants in India, while the modified technology matrix was applied to an IGCC project in China and a fuel cell project in Argentina. The main conclusions of this report can be summarized as follows:

- *Emission baseline estimation is a very difficult and highly uncertain process.* This, in the authors' view, is the most important conclusion to be drawn from the three project analysis. These analysis illustrate that, regardless of the approach, there are potential sources of error. Furthermore, some of these error sources are likely to cause biases in emission reduction estimates at the global level. Of particular importance, the existence of supply-demand imbalances in developing countries is a major complicating factor in project analysis, and a particularly troublesome source of potential errors.

- *Additionality classification errors are of fundamentally greater concern than baseline estimation errors.* This conclusion follows from three main considerations. First, because of asymmetry in the outcomes following upon classification errors, such errors, even if randomly distributed, will lead to the systematic overestimation of market based project emission reductions. Second, because they are more viable than additional projects, the misclassification of non-additional projects as additional will result in preferential investment in these projects at the expense of additional projects. And third, additionality classification errors will always lead to large errors in emission reduction estimates, equal to 100 percent of the estimated project reductions. Rigorous additionality testing may thus provide the best means of guarding against large systematic biases in emission reduction estimates at the global level.

- *To the extent that cost-effective error reduction techniques can be applied, they can be utilized to reduce the potential for systematic errors in the estimation of emission baselines.* However, because it will prove unduly expensive to attempt to eliminate all such errors, and because, in any event, some systematic errors would almost certainly remain even given the most diligent attempts at error elimination, another more cost-effective option may be to explicitly accommodate the likelihood of market mechanism errors in future international agreements. This accommodation may take the form of constant or variable adjustment factors, to be applied to the emission reduction targets specified in future agreements.

- *Since the benchmarking approach to baseline estimation has little direct relevance to the issue of additionality, its application would probably lead to the mis-classification of large numbers of non-additional projects as additional (and vice versa).* Because non-additional projects, by definition, tend to be more viable than additional projects, project developers will exploit the benchmarking approach (knowingly or unknowingly) by preferentially investing in the misclassified non-additional projects, at the expense of truly additional projects. As a result, the number of emission reduction credits awarded will exceed the reductions actually achieved, potentially undermining emission reduction goals.

- *By combining the technology-based test for additionality with the benchmarking approach, the modified technology matrix represents a potential alternative to the benchmarking approach.* The modified technology matrix is applicable to all projects utilizing advanced, non-commercial technologies.

- *Because the modified technology matrix is limited in its scope of application to advanced technologies, it should not be used as an exclusive baseline methodology but should be used in combination with another approach.* For example, the project-specific approach could be applied to projects utilizing conventional commercial technologies not covered by the technology matrix.

1. INTRODUCTION

1.1 Background: Market Mechanisms

Concern about increasing atmospheric concentrations of carbon dioxide and other greenhouse gases, and the potential impact of these increases on the earth's climate, has grown significantly over the past decade. This concern has led to a series of international meetings and agreements seeking to stabilize atmospheric greenhouse gas concentrations. In 1992, at Rio De Janeiro, the Framework Convention on Climate Change was signed by more than 160 countries, including the United States. There was widespread agreement among the signatories on the potential negative effects of climate change under a business-as-usual future. Under the convention, the developed countries (referred to as Annex I countries) were assigned primary responsibility for addressing the climate change issue. However, at the first two Conferences of Parties[2] called to discuss methods for implementing the Framework Convention, there were strong disagreements on what policy instruments should be used to curb global climate change, and what, if any, targets and timetables should be set for achieving emission reductions. Most Annex I nations announced a series of voluntary targets and initiatives for meeting emission reduction goals.

By 1996, it had become clear that greenhouse gas emission levels in most Annex I countries were rising despite voluntary efforts to reduce emissions. A consensus for firmer targets and timetables was building. At the Third Conference of Parties, held in Kyoto, Japan in December 1997, a series of firm emission reduction targets were agreed to by the Parties. Dleveloped countries agreed to reduce their greenhouse gas emissions by an average of 5.2 percent from 1990 levels by 2008-2012. The U.S. agreed to limit its emissions to seven percent below 1990 levels. Since then negotiations on implementing these reductions have stalled and the Protocol has not been ratified. However, major progress in allowing the use of market mechanisms to achieve emission reduction goals occurred at Kyoto. Emissions trading and a new concept where entities can acquire credits for emission reduction activities were among the market-based mechanisms under consideration. This report is concerned exclusively with the latter.

One example of a market mechanism is the Clean Development Mechanism (CDM). The CDM is defined in Article 12 of the Kyoto Protocol as a project activity between a developed country and a developing country that provides the developing country with project financing and technology, while assisting the developed country in meeting its emission reduction commitments. Under the CDM, projects yield emission reductions credits, which are accrued by the developed country and may be applied towards its emission reduction goals. Credits are verified and authenticated units of greenhouse gas reductions from abatement or sequestration projects. They are issued pursuant to the review and certification of each project, by an operational organization (CDM Board) to be

---

[2]The Conference of the Parties (COP) is the supreme body of the United Nations Framework Convention on Climate Change established in 1992. The body meets annually and its primary responsibility is to oversee the implementation of the Convention and the Kyoto Protocol. The Sixth Conference of Parties (COP6) is scheduled for November, 2000.

defined by the Conference of Parties. Discussions at the Fourth Conference of Parties in Buenos Aires divided the project review process into two elements: project certification and project verification. Certification is to take place prior to project implementation, by pre-approved project certifiers qualified to evaluate the emission baselines chosen and the reduction estimation methodology used. Project verification is to be carried out over the life of the project by independent auditors approved by the CDM board. These auditors would confirm that the claimed emission reductions had occurred.

Other examples of market mechanisms include Joint Implementation (defined as an emissions reduction project between two developed countries) and emissions trading. Regardless of which mechanism is utilized, to obtain credits, a market-based emission reduction project will likely need to meet three criteria:

1. voluntary participation,
2. real, measurable, and long-term benefits related to mitigation of climate change, and
3. benefits in addition to any that would occur in the absence of the project activity (referred to as additionality).

Furthermore, market mechanism projects are likely to be required to assist host countries in achieving sustainable development goals. To date international negotiations have not provided specific guidance or criteria for satisfying sustainable development objectives. Rather, current negotiations indicate that individual host governments will be responsible for specifying national sustainable development criteria and approving the sustainability of proposed projects. Thus, a review of a potential project's effectiveness at satisfying national sustainable development objectives are likely to be based on a range of factors including; positive/negative social, economic and environmental impacts; level and quality of technology transfer; contribution to host country goals; emissions reduction efficiency; and project feasibility.[3]

There are multiple approaches for the operation of the market mechanisms and the certification of credits. If these market mechanisms are patterned after the early Activities Implemented Jointly (AIJ) Pilot Phase, it will require a demanding project description specifying all aspects of the contract and a case-by-case approval, and the initial projects will most likely be bilateral and negotiated with the help of environmental or other non-governmental organizations and/or private entities that will bring together the interested parties and help prepare the project documentation required for project approval. A more streamlined approach would include whole classes of projects pre-approved with independent verification to follow. This approach would reduce the barriers to private sector participation.

---

[3]Kyle Graham and Rajini Ramakrishnan. "Sustainable Development, Emissions Reduction Initiatives and the Clean Development Mechanism." Prepared by the Yale Environmental Law Clinic for the Center for Sustainable Development in the Americas, April 1999.

"If the process for estimating emission reductions under the market mechanisms falls prey to political manipulation, gaming, or unintended estimation biases, the cost benefits to be realized through the market mechanisms could be diminished or, perversely, they could even act to undermine and subvert emission reduction goals."

An alternative to the bilateral approach is a multilateral or portfolio approach. This approach has been advocated by representatives of the least developed nations, particularly in Africa, who fear that private sector decision making would exclude their countries from the program altogether. Under this option, a non-Annex I country would offer emission reduction projects to be purchased by the highest bidder. Developing countries could issue certificates and submit them for placement on the market. As the number of projects increase and quantification and verification protocols become more standardized, a domestic and international electronic exchange may develop.

1.1.1 Implications of the Market Mechanisms

One of the potential benefits of market mechanisms is that they will be conducive to the international transfer of new emissions-reducing technologies. They will allow private entities and governments to transfer energy-efficient and environmentally friendly technologies to countries who are not yet responsible for emission reduction targets. It thereby enables a transfer of technologies to countries that may have the largest emission reduction potentials but are not yet able to commit to reducing emissions themselves.

The provisions of real, measurable, and certifiable reductions; voluntary participation; assistance in sustainable development; and significant potential for mitigation are quite challenging for investors to meet. Initially, projects are likely to be restricted to well-defined project types with strong additionality features. In this setting, only those projects with working quantification and verification protocols will be eligible for early approval. The most likely candidates for eligibility fall under the electricity supply, industrial end-use and process efficiency, and low-emitting fuels and vehicle categories.

It is important to recognize that the purpose of market mechanisms is to increase the flexibility and reduce the *costs* associated with meeting emission reduction targets. Market mechanisms are *not* designed to reduce global greenhouse gas emissions beyond emission reduction targets, such as those specified in the Kyoto Protocol. Specifically, developed countries will be able to undertake emission reduction activities outside their borders, while still receiving credit for these activities towards their emission reduction targets. The CDM, for example, would provide for a one-to-one trade between developed and developing countries: for every ton of carbon dioxide emissions reduced in a developing country, a developed country will be allowed to emit an extra ton. As participation in market mechanisms will be driven by the targets imposed on emitters in Annex-1 countries, market mechanism projects are not expected to yield any net change in global emissions. Although participation in the market mechanisms is not by definition required to end as soon as targets have been met, the lack of investment in joint implementation projects during the AIJ Pilot Phase indicates

3

that most private investors will refrain from partaking in joint emission reduction activities unless specific targets or crediting systems are imposed.[4] Thus, investment in market mechanism projects is likely to cease once specific targets imposed by future international greenhouse gas reduction agreements have been met.

The basic assumption underlying the market mechanisms is that the cost of reducing emissions will tend to be lower in some countries than in others. The CDM, for example, was designed to leverage the low-cost emission reduction opportunities in developing countries, while at the same time providing those countries with a means to attract capital for sustainable development projects. *If* market mechanisms can be implemented as originally envisioned, they will be a "win-win" proposition for all involved countries, and it may be determined that the market mechanisms should be allowed to continue even after reduction targets from some future international agreement to reduce greenhouse gas emissions have been met. If, on the other hand, the process for estimating emission reductions under the market mechanisms falls prey to political manipulation, gaming, or unintended estimation biases, the cost benefits to be realized through the market mechanisms could be diminished or, perversely, they could even act to undermine and subvert emission reduction goals.

While from the perspective of Annex I countries, the purpose of the market mechanisms is to reduce the costs associated with reducing emissions, it is crucial to recognize that the market mechanisms may be perceived differently by the developing countries (non-Annex I countries). The latter may view market mechanisms as a means of fostering their sustainable development goals. Under the market mechanisms, developing countries will have access to favorable financing for certain types of projects; market-based projects would be required to not only reduce emissions, but also contribute to the sustainable development goals of the host countries. Sustainable development benefits are not limited to, but should include the following: income benefits, employment, benefits to the local environment, and contribution the local capacity to sustain and build on the projects. Investments in energy efficiency seem to be the most promising:

- they have the highest probability of meeting the criteria of real, measurable, and certifiable emission reductions, as the related emissions can be fairly easily estimated;

- these projects are also the most likely to support sustainable development, as they increase available energy while lowering local and global pollution;

- their potential for significant mitigation and improvement of domestic capacity is also more easily estimated than for most other projects.

Although various formal criteria have been suggested for assessing a project relative to the goal of sustainable development, in the final analysis each host country will judge for itself whether or not

---

[4]"Submission of the United States on the Review of the Activities Implemented Jointly (AIJ) Pilot Phase." February 12, 1999.

proposed projects meet its sustainable development goals.

While market mechanisms in general, may, at least in theory, wed the goal of sustainable development in the developing countries with the goal of reduced costs in the developed countries in a mutually beneficial union, in practice tension may arise between these two objectives. In particular, the developing countries have voiced their concern that they may become a kind of "dumping ground" for high-risk, uneconomic experimental technologies that developed countries are unwilling to develop domestically. Yet, it is precisely advanced, high-risk, marginally economic technologies that most clearly qualify as additional under market mechanisms. Projects utilizing conventional technologies, which may better meet the host countries' goals, may in many cases fail to meet the test for additionality. This underlying tension, and more importantly how it is addressed, may to a large extent decide the success or failure of market mechanisms. Success will require the striking of a delicate balance between developed and developing country goals.

But again, even if successful, market mechanisms will not in and of themselves reduce emissions; they will rather reduce the *costs* associated with emission reduction efforts. It is the emission reduction targets from a future international emission reduction agreements, and not the market mechanisms, that will act as the driving force for reducing global greenhouse gas emissions. Even given a stabilization of developed-country emissions, global emissions, and atmospheric concentrations of carbon dioxide, will continue to rise. Ultimately, the use of market mechanisms may represent only a first step in an international effort to mitigate climate change, and/or to find and adopt strategies for limiting the damage caused by such change.

1.2 Objectives and Report Organization

This report represents a step towards the development of protocols for the estimation of greenhouse gas emission reductions resulting from potential market mechanism projects undertaken in the electric power sector. It deals specifically with the difficult and complex problem of developing emission baselines for carbon offset projects.

Although exchanging credits for emission reduction activities and technologies is a relatively new concept much has already been written about it. The literature has identified and developed a number of approaches to emission baseline estimation under the market mechanism concept, and the pros and cons of each approach have been assessed and reviewed at some length.[5] Much of this ongoing

---

[5]Kenneth M. Chomitz. "Baselines for Greenhouse Gas Reductions: Problems, Precedents, Solution." Draft Paper. World Bank, Washington D.C., July 1998. Ingo Puhl. "Status of Research on Project Baselines Under the UNFCCC and the Kyoto Protocol." OECD and IEA Information Paper. Paris, October 1998; Tim Hargrave, Ned Helme and Ingo Puhl. "Options for Simplifying Baseline Setting for Joint Implementation and Clean Development Mechanism Projects." Center for Clean Air Policy, Washington D.C. November, 1998; and Jane Ellis. "Experience with Emission Baselines Under the AIJ Pilot Phase." OECD Information Paper. Paris, April, 1999.

discussion has proven to be insightful and useful. However, the literature to a large extent has considered emission baseline estimation only in the abstract. Different estimation approaches have been compared and contrasted, but, to date, few attempts have been made to *apply* these approaches. The main exception has occurred under the AIJ Pilot Phase introduced in 1995 under the

---

Emissions Baseline Development Approaches

Project-specific Approach: involves the tailoring of a separate baseline estimation methodology to each individual project, based on a detailed analysis of the project's defining characteristics.

Modified Technology Matrix Approach: a set of technologies is pre-qualified as additional based on a consideration of their economics and current market penetration; baseline emission rates are stipulated.

Benchmark Approach: a set of stipulated baseline emission rates are provided for different countries, sectors, or sub-sectors.

UNFCCC. AIJ has been implemented through Joint Implementation (JI) emission reduction projects undertaken in one country and supported financially by at least one other country. Because project developers were not provided with standard protocols or guidelines for emission reduction estimation under the JI program, a wide variety of *ad hoc*, project-specific approaches were utilized. It is partly in response to the cost, complexity, and subjectivity of developing JI emission reduction estimates on a project-by-project basis, without standardized protocols, that interest in alternative approaches such as benchmarking has grown. However, while these alternatives have been discussed and assessed in the abstract, they have not themselves been applied to projects, real or hypothetical.

Ultimately, any baseline estimation procedures or protocols must prove themselves, not in the abstract, but when applied to actual projects. It would, therefore, be prudent to test the proposed estimation approaches before they must be utilized to assess actual emission reduction projects. The primary goal of this report is to provide case studies of the *application* of alternative estimation approaches to three hypothetical projects. Thus, following an analysis of the three major emission baseline estimation procedures (the project-specific approach, the benchmarking approach, and the technology-based approach) under consideration for market mechanisms, we apply two of these methodologies to hypothetical emission reduction projects.

One project, designed to improve the efficiency of coal-fired power plants in India, is based on an actual project that is sponsored by the U.S. Agency for International Development (USAID), the U.S. Department of Energy's National Energy Technology Laboratory (NETL), the Tennessee Valley Authority (TVA), the Electric Power Research Institute (EPRI), and India's National Thermal Power Corporation (NTPC). This project has yielded measurable efficiency improvements at a number of NTPC power plants, and is being extended both to other NTPC power plants and to coal-fired power plants controlled by India's various State Electricity Boards (SEBs).

Our hypothetical project replicates the activities being performed under the actual India project. Specifically, we assume that these activities are extended to the coal-fired power plants owned by India's SEBs. The hypothetical project is referred to in this report as the Indian Power Plant Efficiency Improvement Project, to distinguish it from the real-world project. In this report, an

The project-specific approach would have been much more difficult and costly to implement than the modified technology matrix approach in the case of the Argentine fuel cell and Chinese IGCC projects, without necessarily yielding more reliable results. However, it would have been impossible to argue for the additionality of the Indian power plant efficiency improvement project without recourse to the project-specific approach.

emissions baseline for the hypothetical project is developed using the project-specific approach also utilized in the pilot AIJ program. The project-specific approach essentially involves the tailoring of a separate baseline estimation methodology to each individual project, based on a detailed analysis of the project's defining characteristics.

The other two projects — an Integrated Gasification Combined Cycle (IGCC) project in China and a fuel cell project in Argentina — are of a more hypothetical nature, although similar projects are being considered by the these and other countries at present. The modified technology matrix approach is utilized to evaluate both of these projects. Under the modified technology matrix approach, a set of technologies is pre-qualified as additional based on a consideration of their economics; stipulated baselines are then provided for each pre-qualifying technology.

Our case studies of the three projects described above involve two distinct steps. First, for each project, we adopt the viewpoint of the project (or technology matrix) developers. From this subjective viewpoint, we attempt to develop as persuasive an argument as possible in support of the additionality of the three projects. In addition, we seek to be as rigorous as possible in our attempts to develop accurate, reliable project baselines and benchmarks.

Then, following the completion of the three project analysis, we adopt an objective viewpoint, and submit the analysis to a critique. By first applying the baseline development approaches in as rigorous a manner as possible, and then subjecting the results of these analysis to an objective critique, our goal is to identify the strengths and weaknesses of each approach under a "best-case" scenario. More specifically, our objective is to identify potential errors sources for each approach, and to provide at least a rough qualitative characterization of these error sources as to their magnitude and potential for biasing global-level emission reduction estimates. Only by approaching baseline development with some rigor can we ensure that the error assessment excludes errors resulting from a mere lack of diligence, and focuses on those more formidable errors that may be inherent to the nature of the estimation approaches, and to the nature of the baseline development problem itself.

Baseline Approaches for Three Sample Projects

Indian Power Plant Efficiency Improvement Project - Project-Specific Approach

IGCC in China - Modified Technology Matrix Approach

Fuel Cells in Argentina - Modified Technology Matrix Approach

---

Our application of two different baseline development approaches — the project-specific approach and the modified technology matrix approach — to the three projects reflects our conviction that each approach has its applications, and that significant benefits are to be gained by combining both approaches in a set of flexible protocols, rather than choosing one over the other for all circumstances. Our project-level analysis bears this contention out: the project-specific approach would have been much more difficult and costly to implement than the modified technology matrix approach in the case of the Argentine fuel cell and Chinese IGCC projects, without necessarily yielding more reliable results, while it would have been impossible to argue for the additionality of the Indian power plant efficiency improvement project without recourse to the project-specific approach. A set of criteria to aid in the selection between the project-specific and modified technology matrix approaches is included in Chapter 3 of this report. In addition to the project-specific and modified technology matrix approaches to baseline development, a third potential approach exists: the benchmark approach. This approach, although not applied to any of the three case studies, is subjected to a separate critique in Chapters 2 and 3.

The organization of the remainder of this report is as follows. Chapter 2 provides an overview of the three main approaches to baseline development that have been proposed: the project-specific approach, the benchmark approach, and the modified technology matrix approach. Specific options under each approach are developed and explored, and the advantages and disadvantages of each approach are discussed. Chapter 3 presents the outlines of the flexible protocol developed for the analysis of the three case studies. This flexible protocol incorporates both the project-specific and modified technology matrix approaches. Chapter 3 also presents a set of criteria for choosing between these two approaches based on project type. Chapters 4, 5 and 6, present the three case studies. The issue of additionality is addressed for each sample project, and the emission baselines for the three projects are quantified. Chapter 7 presents the critique of the three project analysis, including the qualitative error assessment. Finally, Chapter 8 summarizes our main conclusions.

2. APPROACHES FOR QUANTIFYING THE EMISSION BASELINE

2.1 Introduction

In this chapter, we examine the options for developing baselines under market mechanism environment and assess their effectiveness for evaluating power sector projects. We begin by defining the characteristics and criteria for developing greenhouse gas emission baselines. There is a trade-off between the accuracy of the estimated environmental benefits attributed and the level of transaction costs associated with project development. As a result, the options for baseline development are ranged along a continuum where baselines with broad evaluation criteria and low transaction costs are represented at one end, and approaches based on narrow evaluation criteria and high transaction costs are placed at the other end (Figure 2.1).

Figure 2.1. Baseline Option Continuum: Tradeoff Between Transparency, Error, and Transaction Costs

Three options for baseline construction, the project-specific approach, the benchmarking approach, and the modified technology matrix approach will be described and evaluated for their usefulness in

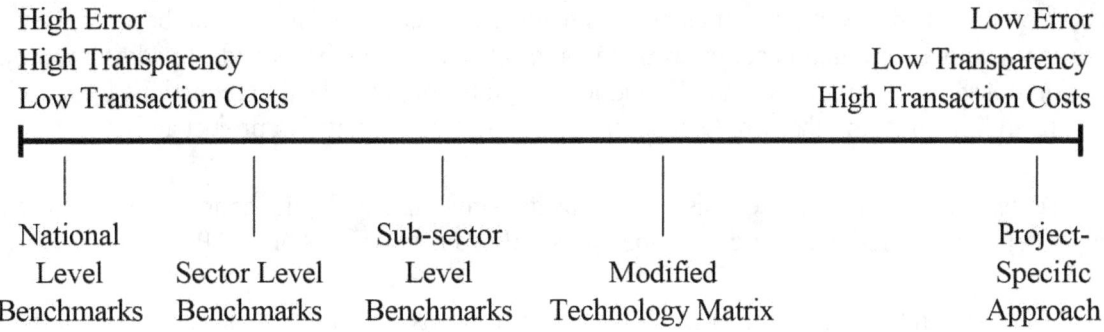

developing power sector projects. Finally, to accommodate the many variations in resources, institutional capacities, and project types across regions and countries, the use of a hybrid approach for project evaluation based on multiple baseline options will be discussed.

2.2 Defining Baselines

Greenhouse gas emission baselines represent the standard from which a measure of valid emission reductions or carbon sequestration is established. To establish the baseline for an individual project, the expected emissions from the business-as-usual scenario (also known as the reference case) are estimated. The baseline can either be based on a forecast of emissions of the actual project that would have been implemented in the absence of the emissions reduction activity, or on a specific set of

9

emission data collected from relevant sectors within the economy. For example, a reference case for a coal-fired power project in a particular host country could be developed based on the emissions of all, or the most recent, coal plants of that specific country. Once the reference case has been constructed, the emissions associated with the proposed project are calculated and subtracted from the emissions of the reference case to determine the actual emission reductions. The calculations involved in baseline construction represent the entire lifetime of the project and include all emissions sources, both primary and secondary, that are relevant to the project.

In the following section, the criteria for developing emission baselines under a tradable emission credits program are described.

2.2.1 Criteria for Baseline Development for Market-Based Emission Reduction Activities

The development of emission baselines that are realistic, verifiable, and accurate is critical to the success of market-based emission reduction activities. Furthermore, the selection of a baseline methodology that facilitates and encourages the implementation of a large number of emission reduction projects is also important. Thus far, little guidance has been provided on how to establish these criteria. For example, the Kyoto Protocol provides only general guidance for developing and certifying emission baselines. According to the Protocol, emission baselines must be set at a level that is both credible and, at the same time, ensures that the project has environmental benefits over and above any climate mitigation activities that would otherwise have happened. In addition, project activities must result in real, measurable, and long-term benefits and lead to reductions in emissions that are additional to any that would occur in the absence of the certified project activity.

These criteria are crucial because emission credits generated in developing countries will be used to help meet emission reduction targets of the industrialized nations, or Annex I Parties.

On the other hand, a large number of carbon offset projects will provide an important source for technology transfer and investment in energy sector improvements in developing countries, thus promoting sustainable development in non-Annex 1 countries. Moreover, increased participation will promote opportunities for developed countries to reduce their emissions in a cost-effective manner.

To ensure that both objectives of environmental integrity and enhanced participation are satisfied, four general requirements must be addressed and incorporated into the baseline methodology applied. These requirements include: demonstration of project additionality, accuracy of the credits generated, high transparency of the certification process, and finally, low transaction costs associated with project development and certification. The following subsections will address the significance of each particular requirement. As will be illustrated, these requirements represent a tradeoff between ensuring the environmental benefits and accuracy of market-based projects and, on the other hand, encouraging participation through a transparent and less costly baseline methodology. In general, the additionality and accuracy requirements promote environmental integrity while the requirements of high transparency and low transaction costs facilitate increased participation.

2.2.1.1 Additionality. The additionality requirement is particularly important as a means of ensuring the environmental integrity of emission credits. Additionality refers to the issue of whether a greenhouse gas abatement or sequestration project will produce emission benefits in addition to those that would have occurred without the project; that is, to receive credits for a carbon offset project, it will be necessary to demonstrate that the project would not have been undertaken were it not for its emission benefits[6]. These projects will never become reality unless an international agreement enters into force and acquire favorable financing, technology transfer, or other market mechanism-specific assistance.

Activities and projects that help reduce emissions will likely occur without the implementation of an international treaty due to competition, technological change and other

---

Key Baseline Requirements

Additionality: The issue of whether an emission reduction project will produce emission benefits in addition to those that would have occurred without the project. The additionality criteria is crucial as it ensures the environmental integrity of the emission credits.

Level of Error: Refers to the level of accuracy in the estimation of emission scenarios and/or the classification of additional and non-additional activities. A high level of accuracy in the evaluation of emission reduction projects is particularly important because even randomly distributed classification errors lead to biased emission reduction estimates. Moreover, additionality classification errors always lead to emission reduction estimation errors equivalent to 100 percent of estimated project reductions.

Transparency: To ensure replicability and independent verification of emission reductions, transparent measures for defining and evaluating baselines are necessary. Thus, common and consistent baseline methodologies are needed, including explicitly defined information and data assumptions.

Transaction Costs: The transaction costs of developing a project include cost of proposal preparation, responding to technical questions raised during project evaluation, travel

---

unforeseen factors that encourage firms to reduce their carbon intensity. Projections for the extent of these activities that will be implemented in a "business-as-usual" scenario have been accounted for in the baseline against which the global reduction targets would be determined. As a result, such activities cannot be counted as offset projects without increasing emissions above the global target, no matter how much they reduce greenhouse gas emissions. Only additional projects, or those that would not have occurred anyway, should receive credit under market mechanisms. Thus, demonstration of additionality becomes the *de facto* test for determining whether or not a project qualifies for credit. Therefore, procedures for developing emission baselines should include an appropriate method for distinguishing additional from non-additional projects. The inclusion of a proper additionality test is likely to raise the transaction costs involved in baseline development. However, as illustrated in the following subsection, such a test is necessary to ensure environmental

---

[6]This is often referred to as environmental additionality. This report focuses only on environmental additionality and does not discuss issues related to financial additionality, i.e., the requirement that financing should be additional to mechanisms such as overseas development assistance (ODA) and UNFCCC-related financial obligations of Annex I countries. It still remains to be decided whether market mechanism projects should also meet the financial additionality requirement.

integrity and single out the potential, additional projects from the ones that are expected to be installed if an international agreement is not implemented.

2.2.1.2 Level of Error. Accuracy in the estimation of the emission baseline ensures the credibility of the generated emission credits. Two issues have a particular impact on the level of error associated with different baseline methodologies. First, the accuracy of the credits generated is directly influenced by the treatment of additionality as a measure of project certification. Moreover, the consideration of temporal issues in the development and quantification of baselines will determine whether accuracy of the emission credits generated throughout the life of the project is maintained over time.

2.2.1.2.1 Additionality Classification Errors. The most important factor influencing the level of error in emissions credits generated is the treatment of additionality during the process of baseline development. A fundamental dilemma presents itself when assessing potential tests for additionality. This dilemma is illustrated in Figures 2.2 and 2.3, which provide a hypothetical distribution of GHG reduction projects in a market mechanism environment. In the figures, we imagine the potential universe of market-based projects, distributed according to the relative ease or difficulty of demonstrating additionality.

In Figure 2.2, a rigorous test for additionality is applied; only those projects to the left of the dashed line will qualify to receive credits under this test. As shown, many additional projects will be mis-classified as non-additional. In Figure 2.3, the test for additionality is relaxed. As a result, most additional projects now qualify for credits, but a large number of non-additional projects will be mis-classified as additional.

Faced with this dilemma, it might be thought that the best additionality test is one that is not too rigorous and not too relaxed. Such an approach, illustrated in Figure 2.4, will lead to a random distribution of classification errors, in which the number of non-additional projects mis-classified as additional will approximately equal the number of additional projects mis-classified as non-additional.

This approach might be regarded as the ideal, because the classification errors will cancel each other out. But a closer analysis will reveal that the errors will not cancel. In fact, *random* errors in the

The Critical Importance of Additionality

To receive emission reduction credits, a candidate project must generate "reductions in emissions that are additional to any that would occur in the absence of the certified project activity." This additionality requirement is crucially important to maintaining the integrity of the credits generated under market mechanism environment. And yet there is an undercurrent of thought, within at least some circles, that additionality is somehow not a "real" issue. This view has, or rather appears to have, a strong logical underpinning: if a project reduces emissions, why should it matter whether or not it would have been undertaken anyway?

This line of reasoning can be quite persuasive because, although it is fallacious, a close, precise analysis of the implications of additionality is required in order to discern the fallacy. The key to understanding additionality, and its crucial importance, is contained in a single word from the above quotation: *certified*. An additional project is a project that will not occur absent *certification*, and the consequent awarding of credits. It is a project that *requires* the reward of credits to be viable.

This means that a project can provide clear environmental benefits, yet still not meet the test for additionality. For example, a project undertaken to improve a power plant's efficiency clearly will reduce the plant's emissions, but if the project would be undertaken regardless of the awarding of credits it is not additional. Why is the test for additionality more stringent than the simple test of environmental benefit? We can begin to grasp the need for stringency by keeping clearly in mind that *the additionality test determines the awarding of credits*. Credits are emission allowances that enable developed countries to *increase* their emissions above potential emission targets. As such, credits may, if awarded carelessly, subvert rather than support the goal of emission reduction. As long as the projects that receive credits are truly additional, they will have no net effect on global emissions. *But if non-additional projects (projects that would have occurred anyway) are erroneously certified as additional, the mere act of awarding credits to these projects will cause net global emissions to increase over and above what they otherwise would have been.*

An example will serve to demonstrate this point. Suppose that an efficiency improvement project is being considered for a particular power plant in a developing country. Suppose, further, that because the project would involve retrofitting advanced, relatively untried technology, it is considered high risk. For this reason the host country project developers are unwilling to invest in the project at prevailing market interest rates. However, the developers have reached an agreement with investors from the United States, who will invest in the project in exchange for any emissions credits. The agreement is contingent upon certification of the project.

This project is clearly additional. If it is not certified, it will not be undertaken, and hence it will have no effect on global emissions. But even if it is certified and undertaken, it will have no *net* effect on global emissions. This follows from the fact that the project will be awarded emission credits equal to the emission reductions it generates. Hence the project's emission reductions will be offset by an increase in emissions elsewhere in the world, equal to the credits awarded to the project. The act of awarding emission credits to this additional project thus has no effect on global emissions--without the award, the project will not be undertaken; with the award, the project's *net* effect on global emissions is zero. This is a fundamental characteristic of all additional projects: they have no *net* effect on global emissions regardless of whether or not they receive certification. Rather than reducing emissions, they serve the purpose of reducing the *costs* associated with meeting emission reduction targets.

The Critical Importance of Additionality (Continued)

But suppose that the host country developers *were* prepared to proceed with the project without the financial assistance of the US backers: i.e., suppose that the project is non-additional. In this case, if the project is recognized as non-additional and rejected as a certified project, it will still be undertaken, and it will generate emission reductions of an amount "x" (where x represents the difference between the emissions of the original, unmodified power plant and the emissions of the plant with the new retrofitted technology). However, if the project is erroneously classified as additional and receives x credits, then the emission reductions generated by the project will be offset by an equivalent emissions increase elsewhere in the world. In short, if the project does not receive credits, emissions will be reduced by an amount equal to x, but if the project is awarded credits, net reductions will be zero. Thus, the mere act of certifying the project causes global emissions to increase by x amount over and above what they otherwise would have been. The awarding of emissions credits to environmentally-beneficial, but non-additional, projects, will in effect undo the environmental benefits that these projects otherwise would have had; the awarding of emission credits to environmentally-beneficial, *additional* projects will have no effect on emissions because, by definition, these projects would not have been implemented absent the credits.

The classification of a project as non-additional does not imply that the project lacks emission benefits; rather it signifies only that the project does not require the incentive of credits in order to be undertaken. Even without the emission credits incentives, many projects will be undertaken that will have the environmental benefit of reducing emissions. However, the emissions effects of these projects will be outweighed by the effects of population and gross domestic product (GDP) growth, and hence emissions will continue to rise. If these non-additional projects were to receive credits, their beneficial effects would be undone, and global emissions would rise even faster.

The critical importance of the additionality issue thus arises from the fact that emission credits (allowances to emit greenhouse gases) will be awarded based on the disposition of this issue. Because they allow the holder to increase emissions, credits must be handled carefully. Specifically, they should be used solely as an *incentive* to projects that otherwise would not be undertaken. If instead they are awarded to projects that need no such incentive, global emissions will increase beyond what they otherwise would have been, regardless of the emission benefits of the projects. If awarded carefully, credits will support the emission reduction goals by reducing the costs of meeting those goals; however, if mishandled and awarded when inappropriate, emission credits, and the market mechanisms that enable their existence, will subvert the very goals they are designed to support.

classification of projects according to their additionality status will lead directly to *systematic* errors in emission reduction estimates. Why? If a non-additional project is approved as additional, it will be undertaken, and it will be awarded credits. However, if an additional project is mis-classified as non-additional, it will not be undertaken, because by definition an additional project will not be implemented absent the awarding of credits. The resulting "lost opportunities" (Figure 2.4) will drive up the costs of meeting emission reduction goals, but the estimation of total emission reductions will remain unaffected. However, when a non-additional project is misclassified as additional, emission reductions are overestimated, and global reduction efforts may consequently fall short of proposed targets. This asymmetry, arising from the very definition of additionality, ensures that even randomly distributed classification errors will lead to biased emission reduction estimates.

Figure 2.2. Projects Qualifying as Additional Under a Rigorous Additionality Test

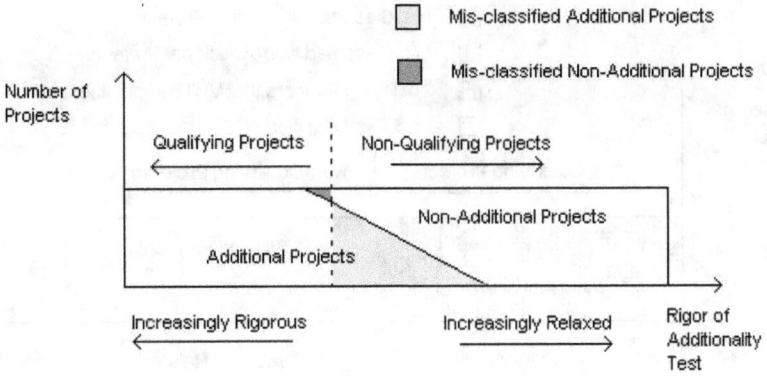

Figure 2.3. Projects Qualifying as Additional Under a Relaxed Additionality Test

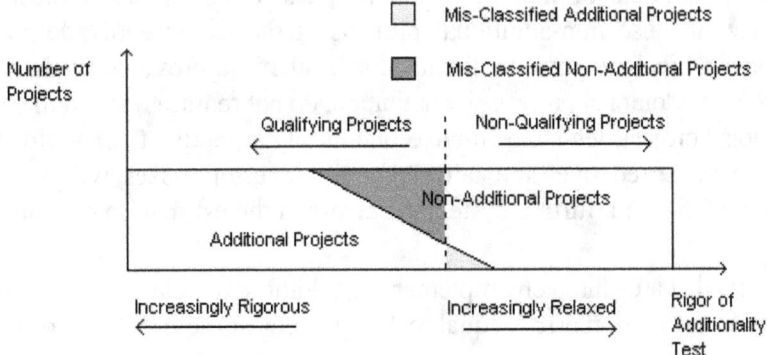

Figure 2.4. Classification Errors and Lost Opportunities Under a "Mid-Range" Additionality Test

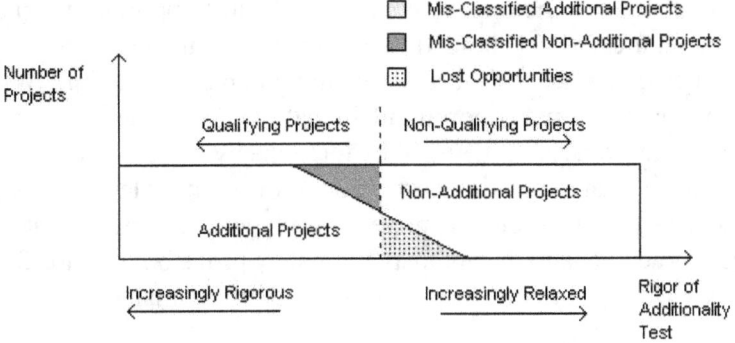

Figure 2.5. Investor Preferences for Qualifying Non-Additional Projects

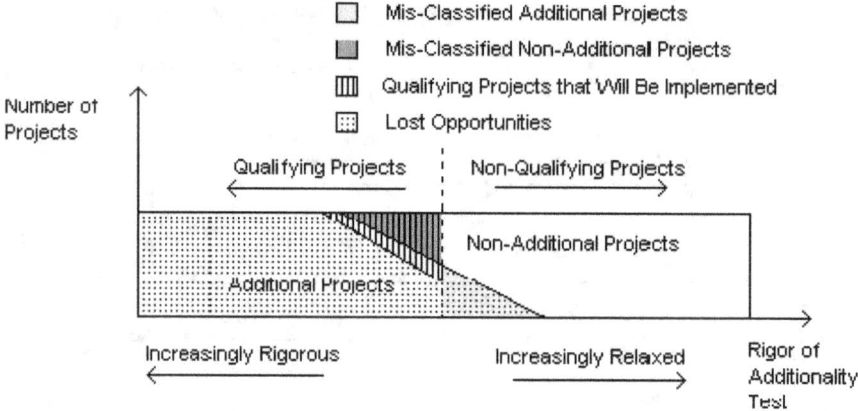

But this is not the only source of biases. Given a relatively "loose" additionality test, that mis-classifies significant numbers of non-additional projects as additional, project developers will preferentially invest in these non-additional projects at the expense of additional projects. By definition, additional projects require the financial and other aid provided by Annex I countries in order to be viable. Non-additional projects, by definition, do not require this aid in order to be viable. In short, non-additional projects tend to be more economically attractive than additional projects, and the former will be preferred over the latter. These investor biases will lead to further lost opportunities (Figure 2.5), and further systematic errors in the estimation of emission reductions.

Finally, for all projects that are ultimately implemented, additionality classification errors *always* lead to emission reduction estimation errors equal to 100 percent of the estimated project reductions.

To summarize, additionality classification errors lead to estimation errors that are highly systematic and very large in magnitude. These estimation errors can be minimized, but only through the application of rigorous additionality tests (as illustrated in Figure 2.6). It should be noted that the costs of rigorous testing, measured in terms of "lost opportunities," should not differ greatly from the costs of "looser" testing. This can be seen by comparing the lost opportunities shown in Figure 2.5 with those shown in Figure 2.6. These figures illustrate that, regardless of the rigor of the additionality test, the "borderline" additional projects falling in the middle third of the diagram are to a large extent lost opportunities. Given rigorous additionality rules, these projects will fail to qualify for crediting; given more relaxed rules, they will be foregone by project developers in favor of non-additional projects. It is true that transaction costs will be lower under less rigorous testing regimes, but these low costs may primarily benefit project developers seeking to qualify non-additional projects. Hence rigorous additionality testing may ultimately prove both cost-effective, and the best means of guarding against large systematic biases in reduction estimates.

2.2.1.2.2 The Treatment of Time. A second factor in the baseline development process that may impact the level of error is the treatment of time. The baseline, against which a market-based project

16

is compared, is purely hypothetical, thus making it difficult to accurately estimate emissions without incurring very large transaction costs. It is particularly challenging to establish whether, at some point during the lifetime of the project counterfactual (the existing project to be replaced), the project developers would have spontaneously switched to a lower carbon intensive activity. Many of the parameters that are necessary to evaluate the expected path of a planned project are hard to observe and are often subject to misrepresentation, strategic manipulation, or unexpected change. However, failure to accurately predict the future emission path of the reference scenario will lead to error in the baseline against which the potential market-based project is compared. As a result, the emission credits assigned will not reflect the true environmental impact of the project. The effect of such error could be significant if a single, standardized emission baseline, such as a benchmark, is used for evaluating entire groups of projects. As a result, flexible procedures that take into account changes occurring over time should be incorporated into the baseline methodology.

Figure 2.6. Use of a Rigorous Additionality Test to Minimize Emission Reduction Estimation Error

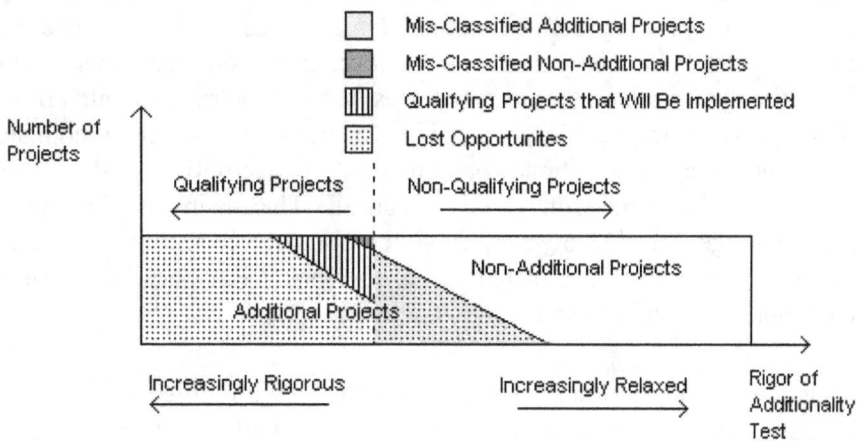

2.2.1.3 Transparency. The level of transparency of a particular baseline approach directly impacts participation in market mechanisms and the credibility of the credits generated. To date, most experience with emission baselines and co-operative abatement activities comes from project activities undertaken as part of the AIJ Pilot Phase introduced in 1995.[7] AIJ has been implemented through Joint Implementation (JI) projects that reduce emissions and enhance carbon sequestration in one

---

[7] In decision 5/CP1, the First Conference of the Parties (COP1) introduced Activities Implemented Jointly under which Parties to the UNFCCC can jointly implement climate mitigation activities. AIJ activities must also meet the additionality requirement. In particular, AIJ financing must be additional to overseas development assistance (ODA) and UNFCCC related financial obligations of Annex I countries. Credits for sequestered or reduced GHG emissions may not accrue to any Party during the Pilot Phase.

country and are supported financially by at least one other country.[8] Each country participating in AIJ has relied on country-specific guidelines for project development and certification, and JI projects have been implemented on an *ad-hoc*, case-by-case or project-specific basis without standardized guidelines for responding to baseline and additionality criteria.

As the AIJ Pilot Phase progressed, the lack of clearly defined assumptions and guidelines regarding methods for quantifying baselines was highlighted as a major problem. It became apparent that even the most rigorous technical analysis of projected emissions had difficulties forecasting future emission scenarios and accurately estimating projected emissions. Because the different AIJ programs did not rely on the same project criteria and assumptions, significant variations in the interpretation of environmental additionality and methodologies for calculating emission baselines were found among similar types of projects within the same region or country.[9] Particularly during the early years of the Pilot Phase, the lack of standardized information requirements hindered project replication, complicated independent verification of projected emission benefits, and may by default have left room for some project sponsors to overstate projected emission reductions.

Thus, to minimize the use of subjective and untestable baseline assumptions for market-based projects, common and explicitly defined methodologies for estimating baseline are needed. The introduction of more clearly defined guidelines would promote increased understanding and prevent complicated and time-consuming approval processes that may delay a potential project to the point where it is no longer economical. Moreover, improved transparency would facilitate replication of projects thereby encouraging additional participation. Objectivity would also be fostered and opportunities for political manipulation would be reduced. This again would increase the credibility of the emission credits generated. However, as will be discussed in section 2.3 of this chapter, there is a tradeoff between the objective of increased transparency and the need for accuracy which must be considered before selecting a baseline methodology.

2.2.1.4 Transaction Costs. The transaction costs associated with project and baseline development will have a considerable impact on the number of projects, particularly smaller sized, that will be applying for credit under the market mechanisms. During the AIJ Pilot Phase, which relied mainly on a project-by-project approach to baseline development and certification, the transaction costs associated with developing a project (including proposal preparation, responding to technical questions raised during the project evaluation process, travel costs, etc.) have been very high,

---

[8] The AIJ Pilot Phase allows all countries who are parties to the UNFCCC to reduce greenhouse gas emissions jointly with other parties. The Joint Implementation program introduced by the Kyoto Protocol will be based on transfer of emission reduction credits among Annex 1 Parties only. This group includes developed countries and countries with economies in transition who have all taken on binding emission reduction targets under the Kyoto Protocol.

[9] U.S. Department of State. "Submission of the United States on the Review of the Activities Implemented Jointly (AIJ) Pilot Phase." February 12, 1999.

reaching a level of about \$100,000 per project regardless of the size of the project.[10] Because many of the transaction costs of project development are fixed regardless of overall project costs, developers of smaller projects have been affected disproportionately. Large-scale projects, such as new power plants, will be less affected by this issue because the transaction costs of project development for a large power plant represent a relatively small share of overall fixed costs.

To ensure market mechanism participation for all types and sizes of projects, the use of standardized approaches to baseline development has been widely promoted. Standardized approaches will reduce the amount of time that individual project developers spend on developing, quantifying, and seeking approval of their emission baseline. In particular, such approaches will reduce the time spent on demonstrating additionality for individual projects. Furthermore, a standardized baseline methodology will be easier to verify and replicate. However, utilization of standardized baseline approaches may have the result that the issues of additionality and accuracy are not adequately addressed, leading to a high degree of error and bias in the estimation of emission credits. The benchmarking approach, which is described in detail in section 2.5 of this chapter, is particularly susceptible to this problem.

2.2.1.5 Summary. To summarize, a demonstration of additionality and consideration of temporal issues must be included in the baseline methodology to prevent error and ensure environmental integrity of the credits generated. However, the ensuing need for stringency in the process of baseline development compromises the goal of reducing transaction costs and increasing transparency. Thus, a trade-off exists between the goal of promoting environmental integrity and enhancing participation in market mechanisms. In the following sections, three baseline methodologies, the project-specific, the benchmarking, and the modified technology matrix approach, will be evaluated with respect to how they respond to this trade-off.

2.3 Review of Baseline Options

To improve on the project-specific baseline approach used during the AIJ Pilot Phase, several new approaches for baseline construction have been proposed with the intent of simplifying the process for developing emission baselines and increasing the environmental integrity of emission credits. To date, discussions of alternative approaches have focused on a variation of the following standardized approaches: benchmarking, modified technology matrix, and top-down baselines.[11] With the goal of

---

[10] Kenneth M. Chomitz. "Baselines for Greenhouse Gas Reductions: Problems, Precedents, Solution." Draft Paper. World Bank, Washington D.C., July 1998. Ingo Puhl. "Status of Research on Project Baselines Under the UNFCCC and the Kyoto Protocol." OECD and IEA Information Paper. Paris, October 1998; Tim Hargrave, Ned Helme and Ingo Puhl. "Options for Simplifying Baseline Setting for Joint Implementation and Clean Development Mechanism Projects." Center for Clean Air Policy, Washington D.C. November, 1998; and Jane Ellis. "Experience with Emission Baselines Under the AIJ Pilot Phase." OECD Information Paper. Paris, April, 1999.

[11] As this study focuses on baseline methodologies for project development, the top-down baseline approach, as proposed by the Center for Clean Air Policy (CCAP), will not be examined further. Top-down baselines are project baselines derived from the binding national emissions target of a host country. The aggregate baseline is allocated

developing baselines for participation in market mechanisms, we will concentrate on only two of the alternative methods, the benchmarking and modified technology matrix approaches, and compare them to the project-specific approach.

As will be illustrated in the following three sections, additionality is assessed differently depending on which methodology a project developer uses. The transaction costs, transparency, and environmental integrity associated with each of the baseline approaches also vary depending on which approach is chosen. As indicated by the different baseline approaches, the need for accuracy often competes with the goal of ensuring low transaction costs and high participation. In this way, baseline options can be placed along a continuum with the broadest and most transparent baseline criteria, and therefore the lowest transaction costs, represented at one end, and the narrowest and least transparent criteria, and therefore the highest transaction costs, placed at the other end of the continuum (Figure 2.1).

At its most general level (i.e., the national level), the benchmarking approach results in low transaction costs but is ineffective distinguishing additional from non-additional projects. This approach should therefore be placed at the end of the continuum that is represented by the broadest baseline criteria. As benchmarks are made less broad and more precise through a division along subsectoral or local/regional lines, they move further towards the middle of the continuum, reducing the level of error when calculating emission reductions, but also increasing transaction costs. The project-specific approach occupies the opposite side of the continuum. Although, it is more technically precise, it usually results in relatively higher transaction costs and less transparency than the benchmark approach. The modified technology matrix approach falls in the middle of this continuum, thus minimizing the trade-off between transparency, error, and transaction costs.

In the following sections, each of the three baseline approaches will be described in detail and the benefits and disadvantages of each method will be evaluated. Moreover, the effectiveness of each baseline approach in terms of evaluating power sector projects will be assessed. (For a summary, see Table 2.1)

2.4 The Project-Specific Approach

The project-specific approach to baseline construction is based on an extensive estimate of total GHG emissions with and without the market-based project activity. If the GHG emissions of the new

---

among individual project activities according to the overall development goals of the country. However, without binding obligations there would be no incentive to enforce compliance with national targets. Non-Annex I countries have not taken on binding national emissions obligations under the Kyoto Protocol. Because the CDM was designed to encourage emission reduction projects in non-Annex I countries, the top-down baseline approach will therefore not be appropriate. Market-based projects using the top-down baseline approach may lead to an unwarranted increase in greenhouse gas emissions in the host country and result in the transfer of offset credits that are not truly additional. Rather, top-down baselines should only be applied to Joint Implementation projects undertaken among Annex I countries because these countries have committed themselves to binding national emission targets.

project are less than that of the project counterfactual, and if the project is proven to be additional to what would otherwise have happened, the project will qualify for emission reduction credits. In this way, the project-specific method allows for a comparison of the intensity of carbon emissions (emissions per unit of output), as well as the level of activity (total project emissions) of the existing and proposed projects.

Using the project-specific approach, the development and approval of project baselines will be dealt with on a case-by-case basis. National programs implementing the AIJ Pilot Phase have relied solely on the project-specific approach for project evaluation. As noted earlier, the project-specific approach has been faulted for its high transaction costs and the level of complexity involved in baseline development. However, as the project-specific approach makes every effort at determining what would have happened in the absence of market mechanism incentives, it is potentially the most accurate method for setting baselines. The development of a standardized set of guidelines for project development and evaluation under this approach will help reduce some of the transaction costs and introduce more consistency and objectivity into the evaluation process.

The following sub-sections examine the methods for assessing additionality under the project-specific approach to baseline evaluation. This includes an evaluation of a project's economic feasibility and an examination of possible non-financial barriers to project implementation. Moreover, the historic, modified, and dynamic options for computing emission baselines will be evaluated.

2.4.1 Additionality and the Project-Specific Approach

In the context of the project-specific approach, additionality is considered as a special issue within the more general issue of baseline development. That is, the question whether the project would have happened in the absence of the project activity is determined separately from the calculation of the baseline. Chronologically, the issue of additionality should be resolved before an attempt is made to quantify the baseline. To do this, a basic question will have to be addressed: does the project *have* a counterfactual? Or, put another way, is the project its own counterfactual? If a project was made possible only by the favorable financing terms accessible through a market mechanism,[12] or through assistance provided for overcoming a pre-specified list of implementation

---

[12]The reference to favorable financial terms assumes that governments, international institutions, and private firms will create financial mechanisms to assist with the development of market-based projects. Mechanisms could include such benefits as the provision of interest free loans, preferential import/export treatment, tax-breaks, technology discounts, etc. Presumably, private project developers will offer such favorable financing at a level no higher than the expected gains from the tradable credits achieved in exchange for emission reductions. Governments, however, may set up additional financial mechanisms to encourage participation in the market mechanisms. It has not yet been determined whether funding should be additional to Global Environment Facility (GEF) funds, official development assistance (ODA) and other developed country financial commitments. However, the economic feasibility test will still be valid even though it is decided that existing development assistance cannot be used in the context of the mechanisms.

Table 2.1. Comparison of Benefits, Uses, Advantages, and Disadvantages of Baseline Approaches

	Project Specific Approach	Benchmark Approach	Modified Technology Matrix Approach
Benefits and Uses	-Larger more capital intensive projects with low transaction costs -Projects that cannot use a standardized baseline approach -Useful for countries or sectors expecting few market-based projects	-Countries with homogenous sectors -Countries expecting large numbers of market-based projects -Useful for evaluating smaller and mid-sized projects	-Countries where volume of market-based projects will be great enough to justify cost of developing the matrix -Useful for evaluating smaller and mid-sized projects
Advantages	-Applicable to all types of projects -More accurate in determining additionality	-Simplifies baseline development -May increase market mechanism participation -Maximizes transparency while minimizing transaction costs	-Environmental integrity is ensured (provides effective additionality screen) -Transaction costs are reduced -Accounts for sustainable development goals of the host country
Disadvantages	-Costs of determining additionality are very high -Lack of transparency leads to increased subjectivity -Currently lacks standardized application procedures and guidelines	-May not distinguish between additional and non-additional projects -High degree of error in assigning emission credits -Fails to distinguish between baseline development and additionality	-Some transaction costs are passed on to the host country and project certification board -Increased chance for political infighting while selecting pre-approved technologies

barriers, then we may conclude that the project would not have occurred but for the market mechanism. Such a project is additional; i.e., it yields emission reductions in addition to those that would have occurred otherwise. If, on the other hand, a project would have been undertaken even without the market mechanism, we may conclude that such a project represents "business as usual." The project would have occurred even without the favorable financing or other benefits made possible under the market mechanism. Hence, its emission reductions are *not* in addition to those that would have occurred anyway. The counterfactual for such a project is the project itself. Once the project has been determined to be non-additional there is no need to proceed to the next issue of quantifying the baseline. Thus, the additionality question is ultimately a question of project motivation: what caused the project? If financial aid or other support obtained via a market mechanism tipped the balance in favor of proceeding with the project, then the project is additional. If the project was undertaken for other reasons (i.e., it was economically viable and institutionally feasible even without the aid), it is not additional.

Of course, it is not easy to determine the motives of a project's sponsors. However, there are procedures that may allow us to infer these motives. Such procedures involve an economic feasibility analysis of the project based on financial/behavioral models and an examination of specific barriers preventing the project from being implemented. The specific techniques for determining additionality using the project-specific approach are described further below (see Figure 2.7).

2.4.1.1 Economic Feasibility Approach. Under the economic feasibility approach, additionality is determined by constructing a financial model describing how the investor would behave over time in the absence of market mechanism incentives and emission credits. If a project is found to be subeconomic without market mechanism participation, it will qualify as additional. To determine market mechanism eligibility, a project submitting a claim for credits would first be assessed for its economic feasibility under the assumption that no credits would be granted. Suppose, for example, that a project's US financial backers plan to provide interest-free loans in exchange for emission reduction credits. In this case, the economic feasibility of the project would be evaluated assuming that the interest-free loans are replaced with loans at the prevailing interest rate. If the analysis indicated that the project would be economically viable without the interest-free loans, then presumably the project would be undertaken irrespective of the awarding of emission reduction credits. Hence, the project would fail to qualify for credits under the additionality criterion. On the other hand, if the project were demonstrated to be subeconomic in the absence of interest-free loans, it would meet the additionality requirement and qualify for emission reduction credits.

The economic feasibility test ensures that projects undertaken do not provide project developers with free emissions credits known as "free carbon." As noted in evaluations of the AIJ Pilot Phase, a majority of the outside parties taking a financial interest in carbon offset projects also claimed some, or all of, the carbon credits associated with each project.[13] In fact, the investors contributed the same amount that they would have invested in a "business as usual" project. As a result, they will receive carbon credits at essentially no cost. However, investors hoping to pass the additionality test under the economic feasibility approach would have to take a larger financial interest than what would have happened in a business as usual scenario to get their project approved. Thus, the problem of "free carbon" would not apply to projects evaluated under this approach.

 2.4.1.1.1 Considerations in the Application of the Economic Feasibility Approach. This subsection summarizes the main considerations necessary in order to apply the economic feasibility test. In particular, we look at the procedures for conducting the economic feasibility test and the complications that may arise from collecting and analyzing the financial data used. A number of decisions related to these issues will have to be made by a certification board and the Parties to the Convention before the economic feasibility test can be effectively applied.

---

[13]"Growth Baselines: Reducing Emissions and Increasing Investment in Developing Countries." Center for Clean Air Policy. Washington, D.C., 1998.

Figure 2.7. Determining Additionality Under the Project Specific Approach

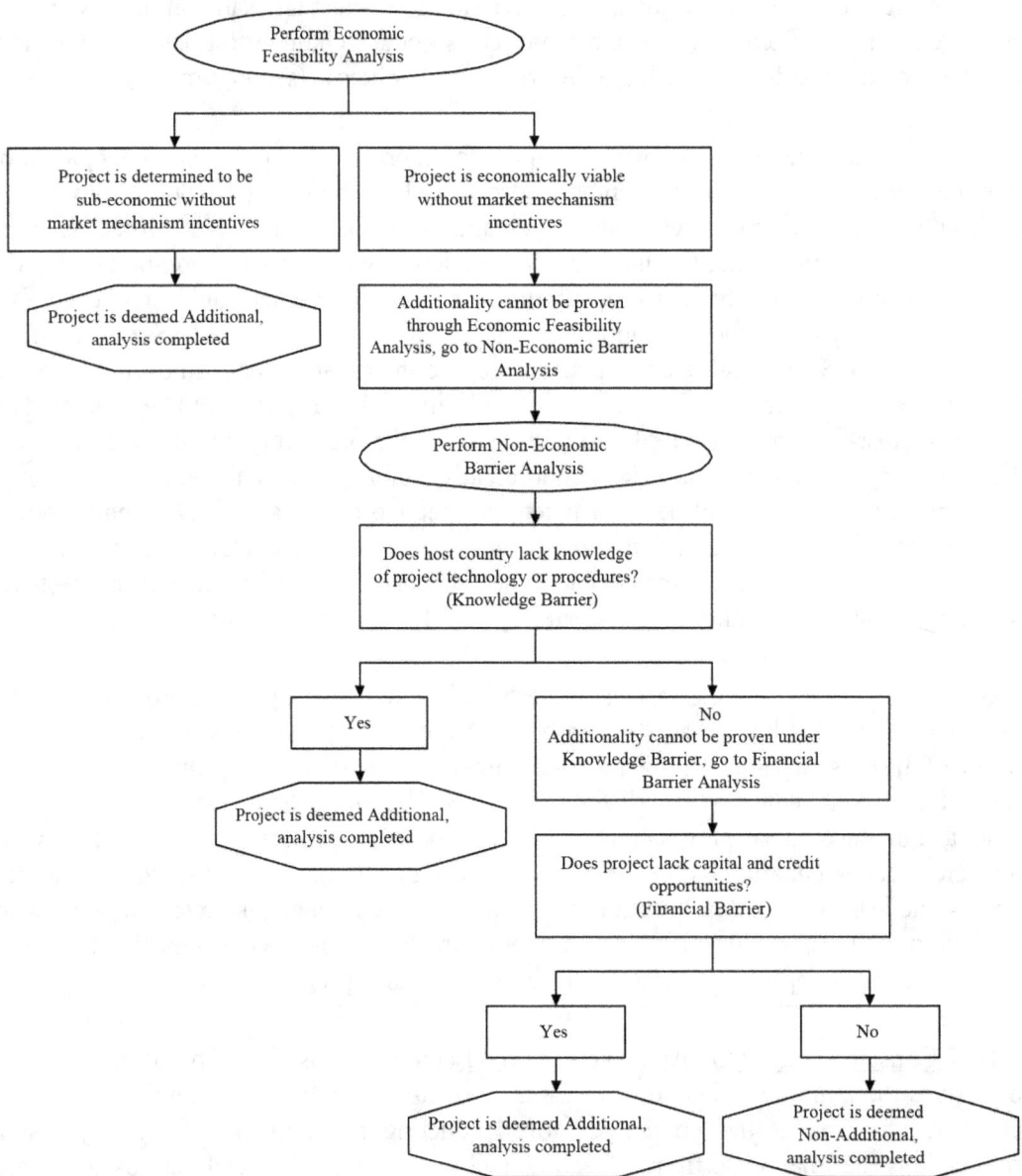

Procedures for Implementing the Economic Feasibility Test

As the first step in assessing a project's claim for emission reduction credits, quantification and verification protocols must include procedures for implementing the economic feasibility analysis described above. Such procedures must be specified in detail, but must also be flexible enough to accommodate the full spectrum of financial instruments (interest-free loans, tax incentives, private or public grants, etc.) likely to be utilized in exchange for emission reduction credits. Specifications to be addressed by a certification board include:

- Factors to determine a project's economic feasibility (internal rate of return, net present value, etc.),

- the computational procedures for computing these factors,

- the allowable sources for the required data inputs, and

- any allowable deviations from the standard approach to accommodate project-specific circumstances.

The most relevant factor is the net present value (NPV) of the project investment. The NPV equals the sum of the up-front investment costs plus the discounted operating costs plus the discounted proceeds. Projects with a positive NPV before the emission credits are taken into account are likely to be commercially viable without market mechanism incentives and, therefore, should not be awarded credits.

Complications in the Use of the Economic Feasibility Test

The use of the economic feasibility approach raises issues that must be addressed to ensure a transparent and fair project certification process.

First, different companies have different economic rate of return hurdles, including differences in the valuation of the discount factor. Before market mechanisms enter into effect, the certification board, therefore, should issue guidelines for evaluating rate of return, including a specification of which factors to include in the evaluation.

A second issue likely to complicate the use of the economic feasibility test is the tendency of private organizations to reject the level of intrusion into their financial planning models that will be required to determine economic feasibility. In cases where some of the data required is unavailable for individual projects, standard default guidelines based on national data, or some other appropriate reference case, will have to be developed on a centralized basis, for example by the certification board. Examples of data to be collected and analyzed include parameters such as:

- capital costs,

- current and anticipated energy prices, and

- pollution charges.

These default parameters would help decrease transaction costs for individual project developers and minimize the temptation to overestimate the amount of emission reductions achieved. Up-front transaction costs, however, will increase the more specific the standard values are made in terms of firm size, type, or location.

2.4.1.2 Non-Economic Barriers. The mere fact that an investment is economically viable in the absence of market mechanism incentives should not exclude the project from inclusion in the program. For many different reasons, projects that are potentially profitable are not always undertaken in real life. Possible reasons include market failures such as poorly functioning capital markets, risks associated with installing and operating locally unknown technology, and institutional barriers or internal organizational structures that discourage investments in energy sector improvements. Even though it can be argued that such factors would be incorporated either in the discount factor or in the expected returns used for the economic analysis of the project, firms do not always include these factors consistently or in similar ways. Consequently, projects that do not pass the economic feasibility test for additionality may also be screened for any non-financial barriers that would prevent the project from being implemented in the absence of market mechanism approval.

In the following subsections, two possible non-economic barriers to project implementation – the knowledge barrier and the access to financing barrier – will be described, and suggestions for how to identify these barriers will be provided.

2.4.1.2.1 The Knowledge Barrier. A possible non-economic barrier to project implementation is lack of knowledge and experience with a particular technology or procedure. Suppose, for example, that a rural solar electrification project to provide lighting in remote regions without adequate access to the electricity grid is commercially viable. However, the project has not been implemented because local knowledge of solar electrification and the procedures for importing, installing, operating and repairing solar technologies is non-existent. Only targeted training and capacity building in local communities and institutions will enable introduction of such a project. Because of this knowledge barrier, project developers could argue that the project is additional. Similarly, it is likely that a group of commercially viable energy-efficiency improvements will not be developed because of a lack of knowledge about the relevant technologies involved. Again, capacity building and training will be necessary to facilitate the introduction of energy-efficiency improvements, and the particular group of energy-efficiency improvements would qualify as additional.

There are no clear-cut and simple procedures for establishing whether lack of knowledge is preventing a specific project from being implemented. However, two indicators may be useful for establishing the presence of a knowledge barrier.

1. *Level of Application of Procedure/Technology.* One method for determining the role of knowledge in project development includes examining whether similar procedures or technologies are used elsewhere in the country or the local area. In the case of an energy-efficiency project for example, the absence of similar initiatives elsewhere in the country is a likely indicator that experience with the particular activity is lacking.

2. *Technical Assistance or Capacity Building Needs.* It may also be possible to show that an activity, such as the solar electrification project, would be implemented once the relevant authorities and users had received training and information regarding the use of photo-voltaic (PV) systems. In this case, it would be possible to argue for the existence of a knowledge barrier.

As experience with market mechanisms grows, other indicators for establishing the presence of a knowledge barrier may be added to the list. The list of possible indicators should be approved and maintained by a central authority, such as a certification board, to ensure standardization of the certification process.

2.4.1.2.2 Lack of Access to Financing. Another possible barrier to project implementation includes lack of capital and credit opportunities. Suppose for example, that the rural solar-electrification project mentioned above is ready to be implemented and the local communities have the necessary experience to operate and repair the PV systems. However, most of the residents lack the funds to make the initial purchase of a PV system because there are no readily accessible credit facilities. As a result, the project cannot be implemented without the creation of appropriate financing mechanisms. In this situation, it would be appropriate to argue that the project is additional as a result of a financing or credit barrier.

A similar financing barrier may prevent large-scale power plants from being implemented. Large power projects are very capital intensive and thus are particularly dependent on the availability of capital. In developing countries, many power projects have failed or have never been implemented because of weak financial markets and limited access to capital. In the Indian power sector for example, capital availability is limited because of a weak financial market structure, restrictions on the use of foreign investment, and government-imposed low electricity tariff rates that restrict the ability of the State Electricity Boards (SEBs) to recoup generation costs. It is therefore extremely difficult for Indian State Governments and the SEBs to raise the capital needed to undertake capacity additions or enter into power purchase agreements with independent power producers (IPPs) – even though the power plants in question are commercially viable. In this case, it may be argued that a proposed power plant is additional because it is impossible to raise the necessary capital to finance the project absent market mechanism incentives.

2.4.1.2.3 Considerations in the Application of the Non-Economic Barriers Test. The use of non-economic barriers to evaluate additionality raises several issues, which will have to be addressed. These include the development of guidelines to apply the non-economic barriers test, equity concerns regarding direction of market-based project investment, and the need for country specific knowledge to properly analyze various non-economic barriers.

Procedures for Implementing and Applying the Non-Economic Barriers Test

Several issues must be addressed by a certification board to ensure a procedure, which is standardized, transparent and easily replicable.

To facilitate baseline development, an official list of possible barriers and the procedures for identifying them could be developed and updated regularly by a certification board as experience with market mechanisms grows. To account for local market and institutional differences, such a list should be developed on a country-by-country basis.

Second, requirements ensuring that projects are made possible only because of the anticipation of market mechanism participation should be outlined. Under the U.S. Initiative on Joint Implementation (USIJI), for example, project developers must prove that the technologies or management practices to be used in the project are an improvement upon prevailing technologies and management practices in terms of their ability to reduce and/or sequester greenhouse gases.[14] Moreover developers must show that the measures to be taken are not required by existing laws or regulations of the host country. However, many developing countries do not have the resources and institutional capacity to enforce laws and regulations. India, for example, has introduced stringent ambient air quality regulations but rarely enforces them. In such cases, it could be argued that a proposed activity, already determined to be additional, should be awarded credit because it has not been implemented for the last five to ten years — even though the proposed activity is already required by existing regulations.

Equity in the Direction of Market Mechanism Investment?

The non-economic barriers that may qualify the project for additionality sometimes result from non-existing or inefficient regulatory policies of the host country. It has been noted that from an equity view-point it is questionable whether governments should be rewarded for failing to introduce effective environmental laws or keeping in place inefficient policies, such as price caps, subsidies, rigid labor laws, and trade restrictions.[15] The fear is that the introduction of market

---

[14] "Resource Document on Project and Proposal Development under the U.S. Initiative on Joint Implementation (USIJI)." U.S. Initiative on Joint Implementation. Version 1.1, Washington D.C., June 1997.

[15] Chomitz, Kenneth M.: "Baselines for Greenhouse Gas Reductions: Problems, Precedents, Solutions." Draft, Carbon Offsets Unit, World Bank. July, 1998.

mechanisms may lead some non-Annex B governments, without binding national emission targets, to prolong inefficient policies in order to attract a larger share of low-cost market-based project investment. Meanwhile, countries that have already removed inefficient policies may receive comparatively less investment and would have to finance additional GHG mitigation activities on their own. A proposed solution includes calculating default parameters for baseline development by using world prices in order to exclude local distortions.

However, this is a hypothetical question that would be difficult to address in the real world, because it would not be possible to identify the various motives underlying domestic policies of individual host countries. The proposed solution of using default parameters based on world prices would seriously undermine the credibility of the generated carbon offsets because the globally derived parameters would overlook natural distortions caused, for example, by geographic differences, variability in access to natural resources, and changes in local market conditions. Moreover, there would be no assurance that the additionality test would be properly applied. Finally, and most importantly, the default approach could be seen as an attempt to interfere with the sovereignty of national governments, and with their right to decide on their own domestic policies. If the default approach came to be viewed as an attempt to impose moral value judgements on developing countries, the long-term viability of market mechanisms could be undermined.

On the other hand, the non-economic barriers test represents a method for evaluating additionality which avoids imposing moral value judgements regarding national policies of the countries involved. The non-economic barriers test is designed to examine the specific facts surrounding individual projects and base an evaluation of possible barriers on these facts alone. The test will be particularly effective once a set of standardized criteria and guidelines for applying the test have been developed by a central authority, such as a certification board.

Knowledge Requirement and Transaction Costs

Finally, it should be noted that the application of the non-economic barriers test will require detailed knowledge regarding political, social and economic factors affecting the specific project — both at a national and a local level. For example, the financial barriers test would require a detailed analysis of various factors, such as capital market structure, capital mobility, and policies on foreign investment that may influence capital availability. Moreover, the use of knowledge barriers as a proof of additionality requires considerable experience with and examination of local markets, social and organizational structures, and institutional procedures. The transaction costs of utilizing this additionality test are therefore likely to be high; both in terms of the effort required of the project developers and that required of the entities responsible for evaluation and certification of the project.

2.4.2 Options for Computing Emission Baselines

Once it has been demonstrated that a project meets the additionality criterion, the next step in assessing the project's emission reductions will be to compute the project's emission baseline. Several different methods for computing baselines exist and these options can be categorized under two broad headings: historical and modified. In this section each of those categories will be addressed and a preliminary assessment of the advantages and disadvantages of each option will be made in the context of developing power sector projects. Finally, the third subsection considers baselines that are dynamic with respect to time, and explains why such dynamic baselines may be required, particularly for long-lived electric generation projects.

> **Emission Baseline Options**
>
> Historical: Defined as emissions in some period prior to the initiation of the project. One advantage is that historical baselines are based on real data rather than projections; however, it is not possible to control for changing conditions.
> Modified: Controls for the effect of factors beyond the project itself. It is an estimate of what emissions would have been "but for the project."
> Dynamic: Modified baseline updated periodically for evaluating long-lived projects.

2.4.2.1 Historical Baselines. Historical baselines are defined as emissions in some period prior to the initiation of the project. There are a number of alternative approaches to developing a historical baseline. For example, actual emissions in the period (year, month, etc.) immediately preceding project start-up might be used as the baseline. Alternatively, emissions in an earlier period might be used if that period were thought to be more representative of typical operating conditions. Similarly, average emissions over a number of consecutive periods might be used to ensure a baseline that represents average or normal conditions.

A key advantage of historical baselines is that they are normally based on data rather than projections or estimates of what emissions would have been in the absence of the project. However, this advantage is often outweighed by a major drawback: it is not possible to control for changing conditions when utilizing a historical baseline. For example, in the case of a project involving an electric generating station, changes in electricity demand or weather can render the comparison between baseline emissions and project emissions meaningless.

There are some types of projects for which conditions remain sufficiently constant to enable the utilization of a historical baseline. In particular, historical baselines are applicable to some types of carbon sequestration projects. Certain simple types of demand-side management (DSM) projects might also be amenable to the application of historic baselines. For example, a historic baseline could be applied to a utility DSM project involving the replacement of refrigerators owned by residential customers with new, super-efficient refrigerators. Because energy consumption of refrigerators remains fairly constant irrespective of changes in the weather or electricity demand as a whole, a historic baseline might provide a reasonable projection of the energy consumption of the old refrigerators.

However, while a historic approach to baseline estimation may yield reliable results for some types of projects, in most cases conditions change too rapidly to allow the use of a historic baseline. For example, while some devices (such as refrigerators) consume energy at a relatively constant rate, many (air conditioners, furnaces, light bulbs, etc.) do not. In particular, historic baselines are almost never applicable to emission reduction projects undertaken at electricity generating plants. The production at such plants fluctuates significantly over time, as a result of a host of non-project-related factors (e.g., weather, population change, economic growth, and outages). Use of a historic baseline could thus lead to significant underestimation of the impact of the project if production increases over time, or a significant overestimation if production declines.

2.4.2.2 Modified Baselines. The purpose of a modified baseline is to control for the effect of factors beyond the project itself. Factors that may have an unexpected effect on the project include inefficient fuel delivery, coal sector strikes, poor maintenance, demand growth, fluctuation in weather patterns, the cost of capital/risk premiums, environmental charges and enforcement levels, maintenance and downtime costs, transaction costs, and public policies affecting prices. A modified baseline is, in essence, an estimate or projection of what emissions would have been "but for the project." A simple example is the unit of production baseline. Here, emissions in some selected base period prior to the project are divided by production (e.g., net generation in the case of a power plant project) during the same period. The resulting fraction is then multiplied by production in the period following project initiation to yield the baseline emissions estimate, as follows:

$$BE_j = (E_i/P_i)P_j, \text{ where}$$

BE_j = Estimated baseline emissions in a period j following project initiation

E_i = Actual emissions in a period I prior to project start up

P_i = Production in period I

P_j = Production in period j

The unit-of-production approach provides a simple method for adjusting baseline emissions to account for fluctuations in such factors as demand growth, weather and other parameters that are hard to predict at the time of project development. It is particularly suited to industries such as the electric power industry, which produce a single, homogenous product (e.g., kilowatt-hours of electricity). However, the unit-of-production method may be less applicable when multiple products are involved, and the mix of products changes over time.

There are many other approaches to developing a modified baseline, ranging from relatively simple engineering algorithms to highly sophisticated models. The key advantage that all of these

31

options offer over historic baselines is that they enable the analyst to isolate the emission reductions that were caused by the project from non-project related reductions. In the context of a system in which emission reduction credits are to be awarded for projects undertaken to reduce emissions, non-project related emissions should be screened out of the emission reduction credits.

2.4.2.3 Dynamic Baselines for Evaluating Long-lived Projects. Power sector projects will typically continue in operation over an extended period of time. However, over the course of decades, an experimental technology may become the conventional technology, and later be replaced as the conventional technology by better technologies. For example, subcritical pulverized-coal combustion technology is being supplanted by state-of-the-art by supercritical technology, which will likely be replaced by IGCC technology when the cost of this technology becomes competitive or the environmental benefits of the technology are required. For this type of long-lived project, the need for dynamic baselines is evident.

One approach to dynamic baseline development would involve comparing a project's emission rate against the average emission rate of comparable power plants with a similar fuel mix. The definition of comparable is, of course, subject to interpretation and should not be applied too narrowly. Certainly, the group selected for the comparison should be representative of the entire power sector. For example, the baseline for an advanced high-efficiency combustion project undertaken at a coal-fired power plant might be based on the average emissions rate of all the coal-fired power plants in the country. The goal is to provide as accurate a representation of the project counterfactual as possible. The average used for the baseline would be updated on a periodic basis, thus ensuring that new technology developments are adequately incorporated in the baseline. Projects implemented in subsequent years would then be derived from the updated baseline alternatives.

Updates will also be required for existing projects that have already been implemented. However, the methodology to be employed for evaluating the baseline for ongoing projects will differ from that used for new projects. The heat rate of a power plant tends to increase as the power plant ages. In addition, major overhauls are undertaken to improve plant availability and efficiency. To account for changes over time in the group of comparable plants selected for developing the baseline, the development of these plants should be traced over time and be used for updating the baseline periodically (e.g., every five years) using the average heat rate for that same group. Thus, the group of power plants originally selected as the basis for the baseline will continue to serve as the baseline throughout the project's life.

The dynamic approach to evaluating long-lived projects provides a guarantee to the project developers that their project will continue to accumulate credits throughout the project. The flexibility of the approach, also ensures that the emission credits earned during the later years of the project are adjusted for overall technological changes in the power sector and in the baseline against which project emission credits are estimated.

32

2.4.3 Evaluation of the Project-Specific Approach

The advantage of the project-specific approach is that it is applicable to all types of projects and is potentially very accurate in terms of distinguishing additional from non-additional projects. However, the transaction costs of project development and determining additionality are often very high, particularly for smaller projects. Moreover, without transparent, consistent and clearly defined guidelines for project development a case-by-case approach can lead to increased subjectivity in project evaluation, thus hindering effective verification and increasing opportunities for disagreement over reduction calculations.

Most of the critique of the project-specific approach has been based on evaluations of the Activities Implemented Jointly (AIJ) Pilot Phase, which suffered from a lack of standardized application procedures and guidelines. However, several steps can be taken to improve the approach. Various evaluations of the AIJ Pilot Phase have proposed a list of modifications to the project-specific approach that would solve some of the problems related to this method. Improvements suggested include increased guidance for evaluating the additionality criteria by applying the economic feasibility test and comparing the proposed activity to a modified baseline. Furthermore, detailed protocols for defining system boundaries, secondary effects, and project-time periods would be beneficial as well. To prevent exploitation of market mechanisms, a system of sanctions could be applied against host countries, project developers, and investigators who purposely falsify project data. Sanctions could rely on monetary fines or on exclusion from participation in the market mechanisms and/or emissions trading. Finally, to promote accurate reporting, a partial-crediting system could be implemented allowing for a margin of error in emissions accounting. Thus, with a more standardized and clearly defined system for evaluating projects, transaction costs are likely to fall and the attractiveness of the project-specific approach will increase.

Once these changes have been incorporated, the project-specific approach will have several advantages. For example, it will be useful for evaluating larger and more capital-intensive projects, where transaction costs represent only a small fraction of total project costs. It will also be effective as a default method for evaluating projects that do not easily lend themselves to a standardized baseline approach. Finally, the project-specific approach represents the most appropriate method for evaluating baselines in countries or sectors expecting only a few market-based projects. In these situations, there will be less opportunity for exploiting the economies of scale otherwise attributed to the more standardized baseline approaches described in the following sections.

2.5 The Benchmarking Approach

The benchmark approach is found at the other end of the baseline continuum, opposite from the project-specific approach as it is likely to result in low transaction costs, high transparency, but also a high level of error. It relies on an average, median, or other metric derived from a defined aggregate

33

or category (such as a specific region, sector, or technology) to determine the amount of emissions reduced by a given project. Based on the performance of this aggregate, a benchmark is then developed, which projects must improve upon, in order to generate valid emission reductions. In contrast to the project-specific approach, which focuses both on the total emissions of the project and the intensity of the project emissions, the benchmark is based on a comparison of emission rates alone. In general, the benchmark will be derived from verifiable information, most likely in the form of the carbon intensity of an output (kilograms of carbon dioxide per kWh) or a product (tons of carbon dioxide per ton of steel). Once the benchmark has been defined, project sponsors only have to show that the carbon intensity of the proposed project falls below the benchmark level to receive offset credits.

The criteria used for developing baselines under the benchmarking approach are very broad, easily replicable, and more transparent than the criteria applied under the project-specific approach. In particular, the elimination of the use of a site-specific, case-by-case estimation of emissions in the absence of the project will increase transparency and reduce transaction costs. Moreover, the costs associated with benchmark development can be shifted away from project developers and host countries towards Annex I countries or relevant international bodies. Thus, participation in market mechanisms is likely to increase. The expectation is that there are economies of scale to be exploited in baseline construction. That is, it is assumed that the added cost of setting the benchmarks will be more than recovered by the avoided cost of developing project-specific baselines. In this way, the success of the benchmarking approach, in terms of lowering transaction costs, is directly related to the number of market-based projects implemented in each benchmark category.

In contrast to the project-specific and modified technology matrix approaches, the benchmarking approach does not address the issue of environmental additionality separately from construction of the emission baseline. Instead all projects that improve on the benchmark automatically receive credit as additional. In other words, additionality almost becomes a non-issue.

Consequently, some amount of projects with low carbon intensities, such as hydro power plants, will receive credit (or a free ride) even though these projects would have happened in the absence of the project activity. This happens because benchmarks represent an average, median or other standard. All projects with lower carbon intensities than the stipulated benchmark will pass the benchmark and qualify for credits. Suppose for example that a benchmark in China is derived from the average carbon intensity of power plants built within the last 5 years. This benchmark would represent a better-than-average standard, assuming that most of these new plants are more efficient that the ones built in previous years. However, most capacity expansions in China involve coal-fired plants. Although better than the average, the benchmark is not stringent enough to prevent natural gas and hydro projects, with much lower or no carbon intensities, from receiving credit. All hydro projects would receive credit – no matter whether they are additional or not.

Similarly, it is possible that a number of activities with carbon intensities above the benchmark threshold will be excluded from market mechanism participation although they would *not* have been implemented absent the market mechanisms. Thus, the benchmarking approach fails to distinguish additional from non-additional projects raising doubts about the environmental integrity of the credits produced.

In the following sub-sections, options for aggregating benchmarks, including national level, sector level, and sub-sector level benchmarks will be examined. In Section 2.5.1 we illustrate that there is a tradeoff between the level of specificity of the benchmarks and the degree of transaction costs involved with aggregating the baselines. In Section 2.5.2, alternative metrics including historical, projected, and normative benchmarks will be examined. Section 2.5.3 assesses the effectiveness of the additionality test as it is applied under the benchmarking approach. Finally, Section 2.5.4 evaluates the benchmarking approach for its usefulness as a baseline estimation method.

2.5.1 Level of Benchmark Aggregation

Several options for aggregating baselines exist (Figure 2.8). These include baselines based on a regional, national, sector, or sub-sector level aggregation. Each of the benchmark approaches represents a different level of tradeoff between transparency, environmental integrity, and transaction costs. As the benchmarks become more specific and disaggregated, they move further towards the middle of the baseline continuum represented by a reduction in the level of error and transparency (Figure 2.1). However, as the

> **Levels of Benchmark Aggregation**
>
> National Level Benchmarks: Measures emissions per capita or emissions per unit of output.
> Sector Level Benchmarks: Measures emissions by industry, such as electricity generation, coal mining, etc.
> Sub-Sector Level Benchmarks: Measures emissions by categories within individual industries.

benchmarks become more specific they also increase the level of transaction costs. Table 2.2 compares the uses, advantages, and disadvantages of each level of benchmark aggregation.

In the following sub-sections, options for developing benchmarks based on sectors, fuel types, technologies, and geographic regions will be discussed and evaluated in terms of their applicability to power sector projects.

2.5.1.1 National Level Benchmarks. National level benchmarks measure emissions per capita or emissions per unit of output. As these criteria are very broad, the national level benchmark approach belongs at the far end of the baseline continuum, furthest from the project-specific approach. National level benchmarks favor investment in non-fossil fuel activities and less energy intensive industries because all industries are subject to the same benchmark threshold. In a very carbon-intensive economy, relying mostly on a single energy source, such as coal, the national level benchmark would favor investment in the least carbon-intensive coal plants as well as any available opportunities for switching to a less carbon-intensive fuel source.

To ensure environmental integrity, the national level benchmark will have to be set at a level that improves on the current emissions intensity; that is, the threshold would be set at a rate below the current average carbon intensity. As a result, many conventional fossil fuel technologies and projects, with carbon intensities higher than the benchmark, would be excluded from market mechanism participation. Likewise, renewable energy projects using hydro, wind, and solar technologies would be more attractive because they have zero carbon intensity and would provide project developers with maximum carbon credit.

Figure 2.8. Different Methods for Establishing the Benchmark

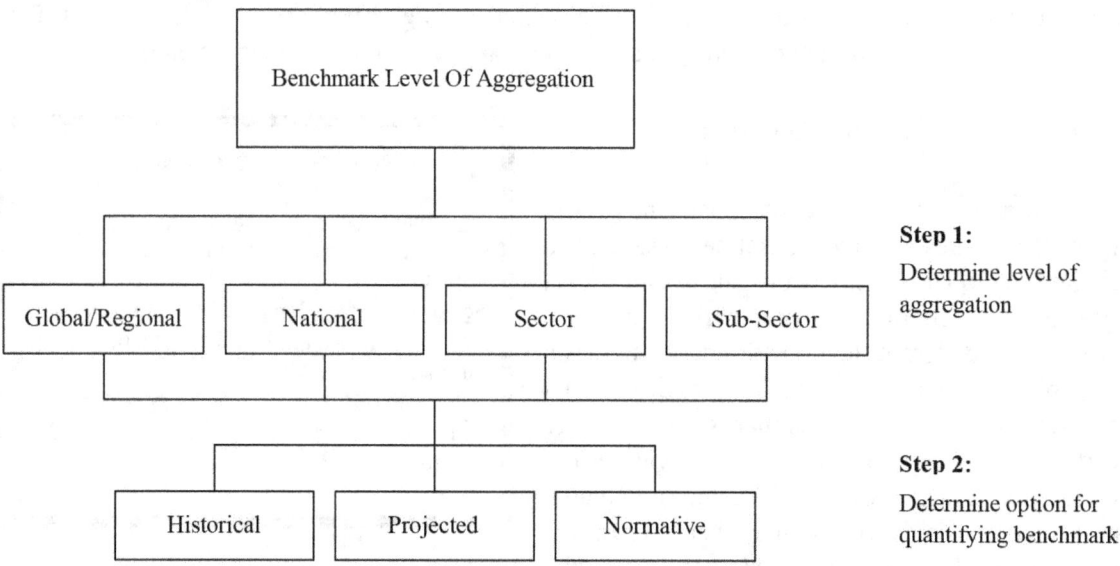

Although this approach may provide significant environmental benefit, the national level benchmark is less likely to serve the sustainable development goals of the host country. In particular, national level benchmarks would provide little incentive for investment in upgrades and improvements to existing fossil fuel power plants that do not represent an improvement compared to the benchmark threshold. Countries that are heavily dependent on coal or oil would be particularly affected by this. India and China, for example, are working to reduce pollution and improve the efficiency of existing fossil fuel plants and they desperately need money to finance such activities. However, national level benchmarks would redirect investment away from these energy intensive industries and discourage initiatives to reduce emissions from existing fossil fuel plants.

2.5.1.2 Sector Level Benchmarks. A straightforward method for benchmark construction entails developing benchmarks for each sector targeted for GHG emission reductions.[16] A common metric

---

[16]Sectors that could be targeted include electricity generation, transmission and distribution of electricity, coalbed methane, transmission and distribution of natural gas, energy-intensive commodity production, and energy consuming equipment.

36

is used for comparing across different fuels, generating technologies, and operation modes. Thus, the sector level benchmark represents a slight move along the baseline continuum toward a more disaggregate approach. Benefits of using aggregate power sector benchmarks include lower development costs and increased incentives for switching to lower-carbon fuels. As another benefit, sector level benchmarks provide a consistent method of crediting different projects across technologies and project types. For example, in the power sector, a sector-wide benchmark developed for new capacity could also be applied to projects that result in electricity savings or additional electricity generation, such as plant retrofits, demand-side efficiency, transmission and distribution (T&D) improvement, off-grid electrification, and distributed generation. By doing this, a level playing field is ensured and no project is favored over another.

Sector level benchmarks are most effective in sectors with a homogenous output. In these sectors, emissions may be measured in terms of unit per output. As project hosts in homogenous sectors are largely unable to alter the nature of the commodity produced, emission reductions are created through real gains in efficiency or through a movement toward less carbon-intensive fuels. The most promising candidate for such benchmarks is the utility sector because each kilowatt hour is equivalent to all others produced. Candidates in the industrial sector include iron, steel, and cement production because of their relatively homogenous outputs.

Some countries may not have the ability to switch to less carbon-intensive fuels and exploit the incentives for fuel switching provided by a sector-wide approach. For example, in countries heavily dependent on coal, such as China, India, and South Africa, it would be largely impossible to shift a significant portion of the electric generating capacity towards natural gas. Furthermore, the long-term cost of clean coal technologies may converge with the cost of natural gas technologies if the cost of natural gas fuel rises and the capital costs of clean coal technologies continue to fall. In this case, benchmarks that discourage coal projects may contradict national sustainability goals and future economic growth. Therefore, the sector-wide benchmarking approach may seriously limit the number of market-based projects implemented.

Table 2.2. Comparison of Uses, Advantages, and Disadvantages of the Levels of Benchmark Aggregation

Level of Aggregation	Uses	Advantages	Disadvantages
National Level Benchmarks	- Measures emissions per capita or emissions per unit of output - Benchmarks are set at a level that improves on current average carbon intensity	- Very low transaction costs - Increases transparency	-High level of error -Favors investment in non-fossil fuel activities and less energy intensive industries -Provides little incentive for investment in upgrades and improvements to existing fossil fuel power plants
Sector Level Benchmarks	-Measures emissions by industry or sector -Uses common metric for comparison across different fuels, generating technologies, and generating modes -Works best in homogenous sectors (e.g., electricity)	-Low transaction costs -Increases transparency -Benchmarks can be applied to new capacity as well as power sector upgrades -Encourages low carbon intensive activities	-High level of error -Some countries (i.e. heavily coal dependent) may not have ability to exploit fuel switching incentives -Does not work well in non-homogenous sectors
Sub-Sector Level Benchmarks	-Measures emissions by categories or fuel sources within individual industries	-Takes into account sectors/industries where fuel switching may be difficult (e.g., best widely used fossil technology for each fuel type) -Reduces level of error somewhat	-Tends to exclude projects based on older more mature technologies -High up-front transaction costs -Useful only in countries with high market mechanism participation
Global/Regional Level Benchmarks	-Measures emissions across national borders as opposed to a country-by-country measurement	-Avoids concentration of investment in a few countries -Increases transparency and lowers transaction costs	-Very high level of error -Limits allocation of investment based on cost-effectiveness

2.5.1.3 Sub-sector Level Benchmarks. Most industrial sectors are highly diverse both in terms of the outputs produced and/or the inputs consumed. The pulp and paper industry, for example, produces numerous outputs that vary in terms of energy intensity (e.g., newsprint, tissue, corrugated cardboard, chipboard, and pulp). The same is true for the chemical industry. In these sectors, switching to less carbon-intensive fuels or production processes may not be possible and it may instead be necessary to devise benchmarks that target specific sub-sectors, fuel types, or technologies. Several types of disaggregate benchmarks have been proposed to account for specific differences in technologies and fuel types within and among countries.

For example, in countries with low-cost domestic oil or coal resources, and a lack of access to natural gas at a reasonable price, a percentile or other benchmark test could be introduced to credit projects utilizing the "best widely used fossil technology" available for a particular fuel source. For example, an IGCC plant in coal-dependent India would be eligible for credit under this benchmark as long as the carbon intensity of the project would fall below the specified benchmark. The "best widely used fossil technology" approach sets a standard for both the fuel type and technology used. However, this approach may exclude some projects that are based on older or more mature technologies – even though such technologies result in low carbon-intensities. As an alternative, benchmarks could be developed for each fuel type used within the economy. Any project capable of improving its carbon intensity below the benchmark level for the specific fuel used would then be eligible for credit. However, a method for evaluating activities that relies on more than one fuel source would have to be devised for this approach to include a maximum number of projects.

A disaggregate benchmark could also be developed to account for different operation modes, such as peakload and baseload. As peaking units often have high carbon intensities compared to baseload units, a sector-wide approach may not create opportunities for crediting improved performance of such units. A parallel benchmark therefore could be set for peakload facilities and for technologies that displace a combination of baseload and peakload power.

Finally, a regional disaggregation of benchmarks at the country level has been proposed, particularly in large countries with highly different energy market conditions. In China, for example, coastal areas have better and cheaper access to imported LPG. Power projects based on natural gas therefore are more likely to be placed in coastal regions while coal-based capacity additions most often will be situated in the interior. Consequently, a less stringent baseline could be developed for areas in China where access to natural gas is limited. However, it should be noted that the process of devising such benchmarks may become extremely politicized as different groups will attempt to advance their particular interests by extending or minimizing the reach of the relevant regions.

Sub-sector benchmarks represent a significant move on the continuum toward a project-specific baseline. Although the benchmarks still do not represent the true counterfactual of what would have happened in the absence of the project, the level of error will be reduced. The precise change in the level of error will be hard to assess because of the difficulties associated with estimating quantitatively the impact of projects that may or may never happen. However, this is an area that would be useful to study in more detail in order to establish the specific effects of various baseline approaches.

Finally, the data requirements for setting benchmarks at a disaggregate level are very demanding and may be difficult to satisfy. The transaction costs of building the institutional capability to develop disaggregate benchmarks are high, and host countries and international bodies under an international agreement are likely to be responsible for these up-front costs. As a result, disaggregate baselines should only be introduced in sectors or countries where the expected volume of market-based projects will be high.

2.5.1.4 Global/Regional Benchmarks. Until now, the discussion of different benchmark methodologies has been based on the assumption that benchmarks will be derived on a country-by-country basis. However, the use of global or regional benchmarks spanning across national borders has also been proposed to avoid concentration of investment in a few countries with less stringent benchmarks. Suppose that as a result of China's reliance on coal, the country's power sector carbon-intensity benchmark is set at a level that is considerably less stringent than the benchmark for Indonesia, which is using comparatively more natural gas. As a result, a developer of a renewable energy project would gain more credits for placing a project in China than in Indonesia – even though the project would have been replacing coal-fired generation in both countries. In this way, countries with the least stringent benchmarks are likely to attract the greatest share of market mechanism investment, because investors will get the highest return (in terms of emission credits per dollar invested) from sponsoring projects in these countries. As many of the Parties involved with the international GHG emissions negotiations have called for the development of a mechanism which encourages an equitable distribution of market mechanism investment among non-Annex I countries, country-specific benchmarks may not be the best solution.

The use of regional or global benchmarks would ensure equity by preventing market mechanism investment from concentrating in a few countries. Small countries, where one large project could drastically change the national baseline, may also prefer to rely on a regional aggregate to ensure equity among projects. The regional or global benchmarks could be aggregated at any of the national, sector, or sub-sector levels discussed above. However, using such benchmarks may increase the level of error in the calculation of emission reductions. Moreover, the regional benchmark may be raised to such a high bar that most of the otherwise additional projects in countries with high carbon intensities and/or fossil fuel dependencies may be excluded from participation.

2.5.2 Options for Computing Benchmarks

Three general approaches have been proposed for quantifying benchmarks and are classified based on the source of information used to derive the benchmarks (Figure 2.3).[17] These include historical, projected, and normative benchmarks.

---

[17] Lazarus, Michael, Sivan Kartha, Michael Ruth, Steve Bernow, and Carolyn Dunmire. "Evaluation of Benchmarking as an Approach for Establishing Clean Development Mechanism Baselines" A report to USEPA prepared by Tellus Institute, Stockholm Environment Institute, and Stratus Consulting, Inc. October 1999.

2.5.2.1 Historical Benchmarks. Historical benchmarks use past information about existing facilities to determine average or median performance of a specific sector. Data used can either focus on the sector as a whole or on information gathered from recent capacity additions. For this approach to work, reliable data must be available regarding key indicators such as amount and type of fuels consumed, physical amount or economic value of product output, and type of process used. A proper metric for the baseline must be defined as well. Options include an average carbon intensity of plants, median carbon intensity, the highest quartile, or perhaps the most efficient plant.

> **Options for Quantifying Benchmarks**
>
> <u>Historical Benchmarks</u>: Use past information about existing facilities to determine average or median performance of a specific sector, focusing on the sector as a whole or on information gathered from recent capacity additions.
> <u>Projected Benchmarks</u>: Based on expectations of future developments and changes to factors like demand growth, availability of capital, etc.
> <u>Normative Benchmarks</u>: Sets baselines that represent an improvement on the average emissions rate to satisfy desired environmental, political, and economic objectives of the benchmarking process.

Developing benchmarks based on historical data has several problems. Studies of recent capacity additions in several countries reveal an increase in the carbon-intensity of many countries. Benchmarks derived from this kind of historical data would thus allow for a less stringent benchmark contradicting the purpose of an international agreement of improving overall emission rates. Furthermore, average historical data cannot account for unpredictable and discontinuous events such as energy price shocks, rapid technological advances, or regulatory change. Finally, the historical benchmarks are data intensive and place considerable demands on the institutional capabilities and resources of the host country. Given the problems with using historical information, this type of baseline should be used only for projects, such as efficiency retrofits at currently operating facilities, where the historical efficiency represents the true counterfactual.

2.5.2.2 Projected Benchmarks. Projected benchmarks are based on expectations of future developments and changes to factors, such as demand growth, market responses to resource prices, capital stock turnover, sector restructuring, availability of capital, and policies relating to greenhouse gas and other environmental objectives. In contrast to historical benchmarks, this approach recognizes that the future may look different from the past. Projections of typical future performance will take into account information, such as planned capacity additions, alterations in market structure, government policies to reduce GHG emissions or accelerate development, availability of capital, and capital stock turnover. Therefore, the projected benchmarks may provide a more accurate explanation of what would have happened in the absence of market mechanism incentives. However, projections of future events are likely to miss significant developments, especially, because developing country economies do not always operate according to model forecasts. Trend lines based on an estimation of projected outcomes therefore are more subjective.

2.5.2.3 Normative Benchmarks. In contrast to the historical and projected benchmark approaches, normative benchmarks do not rely on an estimation of average or median performance. Normative benchmarks set baselines that represent an improvement of the average baselines to satisfy desired environmental, political, and economic objectives of the benchmarking process. For example, a more stringent benchmark will provide incentives for switching towards low carbon-intensive fuels and, by raising the bar, may reduce the number of projects receiving false credits. The normative approach would also allow for adjusting benchmarks across sectors, countries, and regions to avoid leakage and concentration of investment in the countries with the lowest GHG reduction costs. Finally, normative benchmarks could be used in situations of uncertainty and limited information.

Specific types of normative benchmarks include: percentile benchmarks that establish a normative standard for good performance, such as the best five percent, 10 percent, or 25 percent; performance-based benchmarks based on best available technology, efficiency standards, carbon based standards, and consistency with environmental policies; and composite benchmarks that use the above approaches, but include specific modifications, or additionality tests, to be consistent with local constraints and policy objectives. Possible modifications include limiting credits for conventional low-carbon options, such as nuclear and large-scale hydro.

Using the normative benchmarking approach adds a measure of flexibility not available for the historic and projection-based approaches. In particular, the normative approach is useful for ensuring equity by aligning benchmarks with the sustainable development goals of each individual country and for screening out opportunities for obtaining false credits. However, the use of the normative approach as a simulation of the additionality test may be problematic. Indeed, decisions regarding the level of the benchmark will be based on subjective evaluations of what is additional rather than on pre-specified and verifiable guidelines. Moreover, it is not clear what effect normative benchmarks will have on the number of uncredited, but truly additional, projects, because these projects will never be implemented.

2.5.3 Additionality and the Benchmark Approach

A key weakness in the benchmark approach is that it may fail to adequately distinguish additional from non-additional projects. Whereas the benchmarking approach's performance as a baseline development procedure can be improved by going from a sector to a sub-sector analysis, little can be done to improve its performance as an additionality screen. Additionality is a special issue within the more general issue of baseline development and should be resolved as an issue *before* an attempt to quantify the project counterfactual. However, the benchmark approach confounds the issues of additionality and baseline development and fails to deal with the real additionality question of project motivation. This issue will be discussed in more detail in Chapter 3.

2.5.4 Evaluation of the Benchmark Approach

The benchmark approach is useful as it maximizes transparency and minimizes transaction costs. Thus, project participation and foreign direct investment flows to non-Annex I countries are expected to increase. However, owing to the failure of the benchmark approach to effectively distinguish additional from non-additional projects, the level of error may be very high. Moreover, it is anticipated that to maximize project returns project sponsors will preferentially invest in those non-additional (commercially viable) project opportunities, such as hydro projects, that fall below the benchmark.

Furthermore, the expectation of lower transaction costs is predicated on the notion that there is an economy of scale to be exploited in baseline construction. However, the up-front cost of data collection and interpretation for constructing the baseline can be very high, particularly as the benchmarks become more detailed in order to reduce the level of error. Small countries or narrow sectors with only a few opportunities for carbon offset projects therefore may find that the cost of benchmark construction exceeds the potential gains from using this approach.

When assessing the ability of the benchmark approach to reduce transaction costs, it is also important to compare the potential transaction costs of different approaches, not only to each other, but to total project costs. In the power sector, for example, transaction costs are likely to represent only a small percentage of total project costs for large central, power plants regardless of the baseline development procedure utilized. The concern that transaction costs may act as a barrier to projects may thus be overstated, at least for the power sector. Furthermore, the transaction costs associated with a particular baseline development approach should be compared against the cost of the emission reduction estimation errors likely to be introduced through application of the approach. As argued above, the benchmark approach is likely to result in the awarding of credits to a significant number of non-additional projects; as a result true emission reductions may fall significantly short of possible reduction goals.

Finally, the success of the benchmarking approach hinges on the availability of data, particularly for the development of more disaggregated benchmarks, and on the volume of projects undertaken in each sector where a benchmark is established. To minimize error, benchmarks based on a sub-sector analysis are more effective. However, given the lack of detailed information regarding energy sectors and GHG emissions in many potential host countries, the up-front transaction cost of developing the benchmarks will increase markedly. Therefore, the benchmarking approach may only be applicable to a few countries with homogenous sectors expecting a large number of market-based project activities.

2.6 The Modified Technology Matrix Approach

The modified technology matrix approach can be found at the middle of the baseline continuum at the transition point between the more disaggregated benchmarking method and the project specific approach. Following this approach, transaction costs would be kept significantly lower in comparison to the project-specific approach while the degree of error, otherwise associated with the benchmarking approach, would be reduced markedly. In other words, the modified technology matrix represents the middle ground between the objectives of ensuring accuracy and promoting participation in the market mechanisms.

2.6.1 The Early Version of the Technology Matrix

The concept of the technology matrix was originally created as an extension of the benchmarking approach whereby creditable emission reductions of a project are determined through a comparison with a selected group of technologies.[18] Following this approach, the average emissions performance of a number of pre-defined default technologies, which have already reached a predetermined market threshold (defined by a future certification body), would be selected to represent the benchmark for a specific time and within a defined region. In short, a project would be compared to a predetermined matrix of technologies that are readily available locally at the time. Technologies that reduce emissions below the baseline would be considered additional and would automatically receive credit for the amount of emissions reduced. The technology benchmark would be reevaluated regularly and as new technologies reach the market threshold they would be added to the list of technologies on which the benchmark is based. The benchmarks could be differentiated to fit specific technologies, sectors and project types. In sum, the technology matrix is a benchmark derived from current and widely implemented technologies. As a result, any other technologies that improve upon this benchmark would be considered additional.

In essence, this version of the technology matrix approach is very similar to the benchmarking approach already described in section 2.5. As a result, it has several problems, many of which are similar to those of the benchmark approach. In particular, it raises the question of which baseline technologies and performance data to include. For example, a project in Poland could be compared to possible technology mixes as diverse as the average emissions performance of (1) power plants in countries with economies in transition, (2) all power plants in Poland, (3) the most efficient power plants in Poland, (4) all OECD power plants, (5) or finally all commercially feasible power plants.[19] Each choice would result in a different estimate of reductions.

Furthermore, this version of the technology matrix like the benchmarking approach, fails to adequately distinguish additional from non-additional projects. A detailed description of the problems associated with the benchmarking approach is provided in Chapter 3.

2.6.2 A Modified Version of the Technology Matrix

To eliminate some of the problems associated with the above described version of the technology matrix, the authors have modified the procedures for baseline development under the technology matrix approach. We believe that this modification will provide a more effective additionality screen while solving the problem of which technologies to include in the benchmark/baseline.

---

[18] Tim Hargrave, Ned Helme and Ingo Puhl." Options for Simplifying Baseline Setting for Joint Implementation and Clean Development Mechanism Projects," November 1998; "JI Braintrust Group: Minutes of the February 18-19 1998 Meeting," Center for Clean Air Policy; and "JI Braintrust Group: Minutes of the May 4-5 1998 Meeting," Center for Clean Air Policy.

[19] "JI Braintrust Group: Minutes of the February 18-19 1998 Meeting," Center for Clean Air Policy.

Whereas the technologies listed in the original version of the technology matrix in effect represent the project counterfactuals, the technologies listed in the modified technology matrix correspond to those that could be utilized by the project itself. The modified technology matrix, proposed by the authors and illustrated in Table 2.3, consists of a selected list of greenhouse gas abating project technologies that correspond with the sustainable development goals of the host country. The authority to determine whether a given project fits with a country's sustainable development goals rests with the government of the country hosting the project. As sustainable development criteria are likely to vary among countries most examples of the modified technology matrix are anticipated to be country-specific.

Once a list of sustainable technologies has been approved by a host government, stipulated emission baselines are then determined for each technology on the list. However, for a technology to be included in the matrix, it must first be subjected to an additionality test. This test should be based on factors, such as commercial viability and market penetration, and will ensure that non-additional technologies are not included in the list of qualifying technologies. Once it has been proven that a technology is in fact additional, a baseline will be developed for that specific technology based on the emissions performance of a selected group of comparable technologies within that country. Individual projects applying for emissions credit, would then simply demonstrate that the proposed project technology is already listed on the modified technology matrix, and then use the stipulated baseline from the matrix to calculate the emission reductions of the project.

Because of its focus on developing a baseline for each of the technologies on the list, the modified matrix provides a much more accurate mechanism for determining additionality than both the benchmarking approach and the original version of the technology matrix. Indeed, the modified technology matrix combines the technical precision of the project-specific approach with the objectivity and standardization of the benchmarking approach. The stipulated baselines utilized in the modified technology matrix approach are particularly suited to projects involving the opening of new power plants or introduction of new technologies. It is less suited for project involving modifications to existing facilities.

In the following sections, the modified technology matrix will be outlined and methods for evaluating additionality and quantifying the baseline will be described in detail.

2.6.3 Additionality and the Modified Technology Matrix

Unlike the benchmarking approach, which allows for evaluation of several technologies across one or more sectors, the modified technology matrix approach is technology-specific. Thus, a list of default-technologies or procedures is predetermined to be additional (i.e., it is established by the host country and a certification board that a specific technology would not have been introduced in that country or region in the absence of market mechanism incentives). Project sponsors wishing to gain carbon offset credits would then only have to prove that the technology used for their project is already on the pre-defined list of technologies in order to gain credits. To determine the precise amount of credits earned, the expected emissions of the project will then be compared to a baseline stipulated for that specific technology and derived from an analysis of projects in the relevant sector. Any reduction in emissions below the stipulated baseline will receive credit. Table 2.4 provides two examples of this procedure.

The process of determining the additionality of a specific technology would take place in much the same fashion that individual projects are evaluated under the project-specific approach. The evaluation will be based on an examination of the economic feasibility and the market penetration of each particular technology. Figure 2.9 provides an illustration of this process.

2.6.3.1 The Economic Feasibility Test. As a first step in the evaluation of additionality under the modified technology matrix approach, the particular technology is evaluated for its *economic feasibility*. This procedure entails comparing the cost of the specific technology to the cost of alternative technologies in that country to find out if the technology is commercially viable or not.

Besides accounting for the cost of implementing the technology itself, factors to be considered should include energy costs, environmental regulation, cost of capital, demand growth, tariff structures, etc. Other considerations that must be taken into account include whether construction cost of the technology can be predicted with a reasonable certainty and whether the operational performance of the proposed power project can be guaranteed.

If the technology is found to be economic without the favorable financing provided via market mechanism incentives, it will not pass the financial additionality test. However, if the technology proves to be unable to compete with current market technologies in the same sector, it should qualify as additional. Technologies that are likely to pass the financial additionality test include: renewable energy technologies, such as wind, solar, and wave power; fuel cells; integrated gasification combined cycle (IGCC) plants; and integrated gasification fuel cell (IGFC) plants.

2.6.3.2 The Market Penetration Test. Some technologies and procedures may prove to be commercially viable but are still not implemented in selected countries. Non-financial barriers, such as risks associated with installing and operating locally unknown technologies, institutional barriers or internal organizational structures that discourage investment in energy sector improvements, and poorly functioning capital markets may prevent new technologies from being adopted. Therefore, a second additionality test, based on *market penetration*, should also be applied to account for other barriers to implementation that are more difficult to measure and verify. This test should be based on commercially viable projects only and should exclude from the analysis all projects receiving any form of public subsidy or funding.

Table 2.3. Example of a Portion of the Technology Matrix

Countries / Qualifying Technologies	India	China	Argentina	South Africa	Egypt	Philippines	Indonesia	Brazil
Super-Critical Steam Cycle Technology (SCSC)	B_{SCSC-I}	NQ	NA	$B_{SCSC-SA}$	NA	B_{SCSC-P}	$B_{SCSC-In}$	B_{SCSC-B}
Coal-Fired Integrated Gasification Combined Cycle (IGCC)	B_{IGCC-I}	B_{IGCC-C}	NA	$B_{IGCC-SA}$	NA	B_{IGCC-P}	$B_{IGCC-In}$	B_{IGCC-B}
Solid Oxide Fuel Cells (SOFC)	B_{SOFC-I}	B_{SOFC-C}	B_{SOFC-A}	$B_{SOFC-SA}$	B_{SOFC-E}	B_{SOFC-P}	$B_{SOFC-In}$	B_{SOFC-B}
Phosphoric Acid Fuel Cells (PAFC)	B_{PAFC-I}	B_{PAFC-C}	B_{PAFC-A}	$B_{PAFC-SA}$	B_{PAFC-E}	B_{PAFC-P}	$B_{PAFC-In}$	B_{PAFC-B}
Molten Carbonate Fuel Cells (MCFC)	B_{MCFC-I}	B_{MCFC-C}	B_{MCFC-A}	$B_{MCFC-SA}$	B_{MCFC-E}	B_{MCFC-P}	$B_{MCFC-In}$	B_{MCFC-B}
Photovoltaics (PV)	B_{PV-I}	B_{PV-C}	B_{PV-A}	B_{PV-SA}	B_{PV-E}	B_{PV-P}	B_{PV-In}	B_{PV-B}
Pressurized Fluidized Bed Combustion (PFBC)	B_{PFBC-I}	B_{PFBC-C}	NA	$B_{PFBC-SA}$	NA	B_{PFBC-P}	$B_{PFBC-In}$	B_{PFBC-B}

Notes: 1) B = Benchmark value for estimating project baseline emissions.
2) NQ = Not Qualifying. Represents technology choices that do not qualify as additional in a given country.
3) NA = Not Applicable. Represents country/technology combinations that do not fit national sustainable development objectives.
4) This table represents a hypothetical selection of host countries, technologies, and benchmarks that are included mainly for illustrative purposes.

47

Table 2.4. Estimating Emission Reductions Using the Modified Technology Matrix

Application of the Modified Technology Matrix				
Baseline Technology Options	**Baseline Emissions**	**Project Technology**	**Project Emissions**	**Reduction Credits**
Average of most recent coal power plants	X	IGCC	Y	X-Y = Credits
Average of most recent coal power plants	Z	PFBC	V	Z-V = Credits

As available technologies have different market potentials, a certification board and the relevant entities involved with developing the technology matrix should establish a cut-off market penetration rate separately for each technology. Then, if the market penetration of a particular technology turns out to lie below the predetermined cut-off rate, the technology would be found to be additional. Similarly, a certain percentage cap should be instituted as the cap above which initially accepted technologies will be removed from the matrix. Finally, if it turns out that the targeted market penetration rate has been achieved, but that this happened only because of government funding of the technology, then the technology should also qualify as additional.

However, before all this can be done, guidelines will have to be developed by the Conference of the Parties or a certification board for establishing an appropriate market penetration cut-off rate. The rates should be based on a percentage of capacity installed and would have to be set a level that is low enough (such as 1-2 percent) to show that if a particular technology is used in more than a couple of projects within a given country, then the technology no longer qualifies as additional.

2.6.4 Quantifying Stipulated Baselines

A critical issue for the success of the modified technology matrix will be the choice of the emission baseline for each of the additionality-approved technologies on the list. Under the benchmarking approach, emission baselines are derived from sector-wide or sub-sector based energy intensity data, using the metric of kilograms of carbon dioxide emissions per kWh. The stipulated baselines of the modified technology matrix could utilize this same approach but they do not necessarily have to be derived this way. They could, for example, be derived based on the average thermal efficiency of the conventional technology most likely to have been used, but for the project. In this way, factors, such as average heat rate, fuel mix, and best engineering practices, would be included in the baseline. Technologies using a specific fuel type, such as coal, might be compared to projects and technologies using that same fuel. For instance, the specific amount of credits achieved from an IGCC project in China could be derived by comparing the

48

Figure 2.9: Determining Additionality Under the Technology Matrix Approach

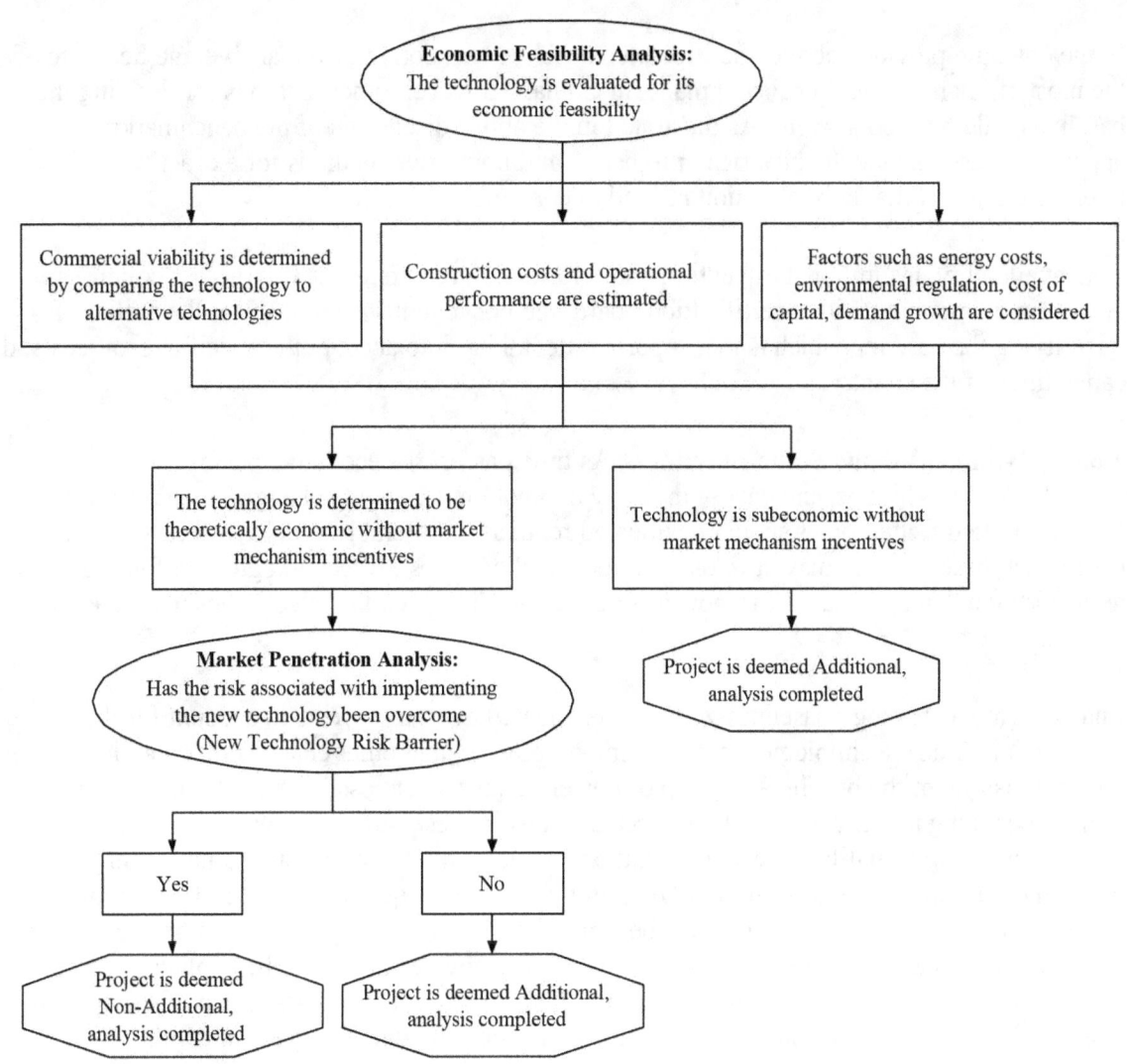

average heat rate of the most efficient conventional coal plants in China against the heat rate for the specific IGCC plant in question. The difference between the two heat rates would then be multiplied by the generation from the proposed IGCC plant to find the specific amount of credits earned. For each technology on the matrix, separate stipulated baselines should be developed for generating units of different sizes and (capacity) ranges.

In the example provided above, the stipulated baseline was derived from an average heat rate of the most efficient conventional coal plants in China. However, other methods for deriving the baseline could be used as well. As illustrated in the above discussion of the benchmarking approach, these include the historical, projected, and normative methods for extracting and interpreting the available information needed to construct a baseline.

Data availability, institutional capacities, and information requirements vary across countries. Hence, in cooperation with a certification board, the host country should select the method for quantifying the baseline which is most appropriate, taking into account the specific resources and capabilities of the country.

2.6.4.1 Dynamic Versus Static Baselines. As time passes, the economic performance, technological capabilities, and energy intensity of a nation is likely to change. As a result, the list of pre-qualified technologies should be updated regularly, preferably every five years, to capture the impact these changes may have on individual technologies. If technologies are found no longer to be additional, they should be removed from the list and added to the activities that make up the baseline.

Likewise, the technology baselines will also be updated every five years to account for the introduction of new technologies, heat rate improvements, and other changes that may influence the composition of the benchmark group of power plants used to establish the baseline. An initial group of existing power plants will be selected as best representing the "typical" counterfactuals for projects using a qualifying technology; the average heat rate or emissions rate for this benchmark group will be applied to the first group of projects qualifying under the technology matrix. However, after five years a new benchmark group, reflecting changes/improvements in power plant technology, will be used as the basis for a new benchmark to be applied to all new projects implemented as of year six In a similar fashion a new benchmark group of existing power plants will be used to establish a new benchmark at each following five-year interval.

If a technology is removed from the matrix, existing market-based projects that introduced this technology prior to the time of removal from the matrix should be allowed to continue to earn credits. Large power projects, in particular, are locked into their investments for a long time and project developers need a guarantee that they will be able to recover the additional costs of undertaking a project that is subeconomic at the time of implementation. Thus, to make sure that project developers will be willing to invest in large-scale capital investment projects, they should be allowed to generate credits for the entire economic life of the project to recover their fixed costs.

This does not mean that the baseline has to remain static. However, the methodology to be employed for updating the benchmarks for ongoing projects will differ from that used for new projects. As power plants often remain in operation for 50 years or more it would be unrealistic to base the stipulated baseline on a benchmark of existing power plants that is updated every five years to incorporate the heat rate of recently constructed plants. Such an approach would be based on the assumption that the counterfactual power plant would be shut down and replaced with a new power plant every five years. However, to account for the changes in the heat rate of the counterfactual power plant, as a result of wear and tear on the equipment, the stipulated baseline should be based on an analysis of a representative group of power plants, of a similar age as the power plant in question. The average heat rate for this group of plants will be used as the benchmark for the first five years of the project's existence. The development of these plants should then be traced over time and be used for updating the benchmark every five years using the average heat rate for the same group. In this way, the power plants and technologies originally selected for developing the stipulated baseline will continue to serve as the benchmark throughout the life of a project, while the benchmark heat rate will vary over time. The application of this dynamic approach to baseline development using the modified technology matrix will be described further in Chapter 5, which describes the quantification of a baseline for an IGCC project in China.

2.6.5 Evaluation of the Modified Technology Matrix Approach

Using the modified technology matrix approach, environmental integrity is improved while the transaction costs to the project developer are reduced. Moreover, the modified technology matrix approach considers the sustainable development goals of the host country directly because only those technologies supporting the goals of the host country will be allowed on the list. However, some of the transaction costs, previously resting with the project developer, will now be passed to the host country institutions and a certification board responsible for developing and approving the technology matrix. Another potential problem associated with the modified technology matrix is the increased opportunity for political infighting during the process of selecting the pre-approved technologies.

Host countries should therefore mostly consider using the modified technology matrix in markets where the expected volume of market mechanism investment will be great enough to justify the initial political and transaction costs of developing the matrix. As part of these considerations, it should also be emphasized that transaction costs cover only a small portion of the overall costs of developing and constructing large power plants. As a result, the modified technology matrix approach may turn out to be most useful for evaluating smaller and mid-size market-based projects.

2.7 Conclusion

The above discussion of the three baseline methodologies illustrates the range of options available, spanning from the broadest and most transparent criteria to the narrowest and most subjective criteria. Each option represents a tradeoff between the goals of increasing transparency, reducing error, and minimizing transaction costs. The choice of a baseline option will have different effects depending on the size and type of the project in question. For example, projects involving the opening of new power plants are better suited to the stipulated baselines utilized in the modified technology matrix approach. Meanwhile, projects designed to improve existing plants that cannot be evaluated on the basis of technology alone will obtain a more accurate estimate of emission reductions by using the project-specific approach. Finally, the benchmark approach should only be applied to sectors with homogenous outputs and many opportunities for market-based project activities.

Chapter 3 outlines a generic procedure for selecting an appropriate baseline methodology and applying it to a specific project.

3. GENERIC PROCEDURE FOR ESTABLISHING EMISSION BASELINES

In the preceding chapter the principal proposed procedures for developing project emission baselines were reviewed, and the advantages and disadvantages of each procedure were assessed. In this chapter, the preceding review will be used as the basis for developing a generic approach to baseline estimation. Then, in Chapters 4, 5, and 6, we will test and illustrate this generic approach by applying it to three hypothetical projects.

3.1 Assessment of the Three Proposed Procedures

Two primary conclusions are drawn based on our assessment of the three proposed procedures. First, for purposes of our case study analysis, the procedures employed should be limited to the project-specific approach and the modified technology matrix approach. The focus of this report is on identifying and reducing potential errors associated with emission baseline development. As discussed in Chapter 2, the project-specific and modified technology matrix approaches should in general be less prone to error than the benchmark approach. This is not to say that the benchmark approach does not offer significant advantages, especially in terms of transaction cost reductions. However, as estimating transaction costs is not a goal of this report, we want to keep the focus of the case studies on error rather than cost reduction. Rather than treating benchmarking through a detailed case study approach, we will instead provide an overview of error sources associated with benchmarking in the next subsection.

Second, the choice between the project-specific approach and the modified technology matrix approach is best made on a project-by-project basis, because each approach offers significant advantages over the other depending on the specific circumstances. In subsection 3.1.2 the specific areas of applicability of each of these two approaches will be further developed.

3.1.1 Potential Error Sources in The Benchmarking Approach

Without doubt the benchmarking approach to baseline development offers significant advantages in terms of cost reduction, transparency, and objectivity. It is for these reasons that it has garnered considerable attention in recent months, as a potential alternative to the much more costly project-specific approach. However, it is important to recognize that the benefits of the benchmarking approach may come at a significant cost. As discussed in Chapter 2, the tradeoff is between cost reduction, on the one hand, and accuracy on the other. Recent discussion of the benchmarking approach has perhaps focused more on the cost reduction benefits, without fully exploring the error implications. Therefore, in this subsection we offer an assessment of the potential for errors in the benchmarking approach, and in particular in its treatment of additionality. Our purpose here is not to prejudge the usefulness of this approach. It may well be that, when all is said and done, the cost benefits associated with benchmarking will be judged to outweigh the error implications. Rather, our goal is merely to provide a full assessment of the potential for errors, so that the cost/error tradeoff can be made on the basis of an informed judgement.

> **Critique of the Benchmark Approach to Additionality**
>
> Under the benchmark approach, the project's emission rate is compared with a pre-selected benchmark. If the project's emission rate is higher than the benchmark, the project is deemed non-additional; if it is lower, the project qualifies as additional. This simple numeric comparison of two emission rates does not address the key question determining additionality: Is the project viable absent market mechanism incentives? For this reason it is anticipated that, if used, the benchmark approach will result in the misclassification of a large number of non-additional projects as additional (and vice versa).
>
> Furthermore, it is anticipated that project developers seeking emission reduction credits will preferentially invest in these misclassified non-additional projects, at the expense of additional projects. This follows from the fact that the latter projects will tend to be less viable than the former projects, by definition.

As was discussed in the preceding chapter, the benchmarking approach offers project sponsors significant opportunities for gaining emission reduction credits without reducing emissions. In fact, the benchmark approach strongly favors investment in non-additional projects at the expense of additional projects. First, non-additional projects will tend to be more economically viable than additional projects. By definition, additional projects require the financial and other aid provided by the developed country sponsors to be economically feasible or to overcome the financial and other barriers to project implementation. Non-additional projects, by definition, do not require the participation of the developed country sponsors in order to be viable, and in fact they will be undertaken with or without this participation. Therefore, given a baseline estimation procedure that erroneously pre-qualifies a significant number of non-additional projects as additional, developed country sponsors will tend to invest in these non-additional projects rather than truly additional projects. Second, a numeric benchmark will, at best, prove a very crude screen for additionality. Whatever power the benchmark approach possesses to distinguish additional from non-additional projects is largely coincidental. A comparison of a project's emission rate with a preselected benchmark value has little direct relevance to the question of the project's additionality.[20] The benchmarking approach does not really address the issue of additionality, and for this reason it will result in the mis-classification of many non-additional projects as additional (and vice versa). To understand why this is the case, it is necessary to understand the nature of the issues underlying additionality.

For any particular project, before making an attempt to quantify the project counterfactual, it is first necessary to address a more basic question: does the project *have* a counterfactual? Or, put another way, is the project its own counterfactual? If a project was made possible only by the favorable financing terms gained in exchange for emission reduction credits, then we may conclude that it would not have occurred but for the market mechanism incentives. Such a project is additional; i.e., it yields

---

[20]This is particularly true for benchmarks representing a mean or median emissions rate for a particular sector or subsector. Benchmarks representing the best (i.e., lowest-emitting) units in a sector can, it may be argued, provide some ability to distinguish business-as-usual from additional projects. Nonetheless, even the latter type of benchmark does not directly address the question of project motivation, which is the key question underlying the issue of additionality.

emission reductions in addition to those that otherwise would have occurred. If, on the other hand, a project would have been undertaken even without market mechanism incentives, we may conclude that such a project represents "business as usual;" it would have occurred even without favorable financing and hence its emission reductions are *not* in addition to those that would have occurred anyway. The counterfactual for such a project is the project itself; once the project has been determined to be non-additional there is no need to proceed to the next issue of quantifying the baseline.

The additionality question is ultimately a question of project motivation: what caused the project sponsors to undertake the project? If the financial aid obtained via a market mechanism tipped the balance in favor of proceeding with the project, then the project is additional; if the project was undertaken for other reasons (e.g., it was economically viable even without the aid) it is not additional.

It is virtually impossible to determine the motives of a project's sponsors. Fortunately, there are procedures that may allow us to infer these motives. One such approach would involve an economic feasibility analysis of the project, under the assumption that the project receives financing at the prevailing market conditions. If the project is found to be sub-economic without the favorable financing obtainable via a market mechanism, then it might reasonably be concluded that the project sponsors undertook the project because of the favorable financing. Such an economic feasibility analysis approach thus may provide some insight into the project's motivations.

The benchmark approach does not address the question of project motivation. Under the benchmark approach, all projects with an emissions rate above a benchmark value are deemed non-additional, while all projects with a rate below the benchmark are judged additional. This simple numeric comparison of emission rates does not offer any insight into the motives of the project's sponsors.

Because the benchmark approach represents an attempt to address additionality through a purely numeric or mathematical comparison of emission rates, it is appropriate to assess it from a mathematical viewpoint. When stated mathematically, two separate questions must be addressed in evaluating a project for credit:

1. Is the counterfactual emissions rate equal to the project emissions rate (the additionality question)?

2. If not, what is the counterfactual emissions rate (the baseline development question)?

Both of the above questions must always be understood to apply to a specific individual project under evaluation. Nonetheless, the second question can (in fact must) be answered with an approximation. Hence, although the procedure for developing a benchmark is entirely divorced from the specific project to be evaluated, the benchmark may nonetheless be used as an *approximation* to the project's counterfactual emissions rate, particularly if the benchmark was derived from similar projects. However, the first question is an equivalency question that does not admit the use of an approximation; either the project and the counterfactual are the same, or they are not. A benchmark can never serve as anything more than an approximation to the project's counterfactual; a comparison of an *approximate* counterfactual with a project emissions rate cannot establish the equivalency of the latter with the *true* counterfactual. Stated mathematically, if p is the project emissions rate, B is the benchmark emissions rate which is approximately equal to tc (the true counterfactual emissions rate), and p>B, it clearly does *not* follow that p=tc. Yet under the benchmark approach, the conclusion is that p *does* equal tc; i.e., the project emissions rate is the same as the counterfactual emissions rate, and hence the project is not additional. Similarly, if p<B, it does not follow that p^=tc, although this would be our conclusion under the benchmark approach.[21] An appropriate use of a benchmark is as an approximation to a project counterfactual, but such a benchmark, derived through a sectoral analysis entirely divorced from any consideration of the project at hand, has little relevance to the question of that project's additionality.

Because the benchmark approach does not directly address the question of a project's additionality, it will almost certainly result in the mis-classification of a large number of non-additional projects as additional, and vice versa. For example, in the power sector, all projects involving new and existing hydroelectric and nuclear facilities will qualify as additional, without even considering the question of whether or not they represent business as usual. Depending on how the benchmark is set, many natural gas projects will also qualify as additional, regardless of their viability absent the market mechanism incentives. On the other hand, coal-fired projects involving advanced technologies, such as IGCC, will probably be classified as non-additional, even though such technologies are not commercially viable at present. Given the large classification errors likely to result, coupled with the economic advantages of non-additional over additional projects, the benchmarking approach will probably lead to a strong bias towards investment in non-additional projects, the awarding of a large number of false emission reduction credits, and a consequent shortfall in efforts to meet emission reduction goals.

3.1.2 Applicability of the Project-Specific and Technology Matrix Approaches

Unlike the benchmark approach, both the project-specific and modified technology matrix approaches are designed to directly address the crucial issue of additionality. In the case of the project-specific approach, the viability of each individual project, without market mechanism incentives, is assessed using such means as economic feasibility analysis and project barrier analysis. And although the modified technology matrix approach deals with entire technologies rather than individual projects,

---

[21]Oddly enough, as this analysis shows, the use of the benchmark to establish additionality is not even logically consistent with its use as an approximation to the counterfactual. Logical consistency would require that the benchmark be compared to the project emissions rate, and if found to be *equal*, or perhaps nearly equal (rather than *larger*), the project would be judged to be non-additional. If the benchmark were not equal (or not close) to the project emissions rate, the project would be deemed additional. Of course, such an approach would still not be mathematically valid, because it involves the use of an approximation to test for equivalency.

it nonetheless involves a direct assessment of the commercial viability of these technologies, and in this way it addresses the issue of additionality directly and explicitly.

As was discussed in Chapter 2, both approaches offer advantages and disadvantages. When implemented in a careful, rigorous manner, the project-specific approach is likely to yield a more reliable estimate of the project counterfactual. However, because it must be applied on a project-by-project basis, and because it requires a projection of what would have happened "but for the project," the project-specific approach will in many cases prove to be difficult, complex, and costly to implement. Also, because the project-specific approach is designed to account for the specific conditions and circumstances surrounding each individual project, it is not readily amenable to

> **Key Point**
> The project specific approach and the modified technology matrix approach both directly address the issue of additionality, but in different ways. Under the project-specific approach, the viability of each project without the market mechanism incentives is assessed using economic feasibility analysis techniques or project barrier analysis. Under the modified technology matrix, additionality is addressed through direct assessment of the commercial viability of each technology.

standardization. However, some limited degree of standardization is possible, at least in terms of the basic approach to evaluating a project.

The modified technology matrix approach, on the other hand, eliminates the need to perform a project-level assessment of the additionality issue, since it pre-qualifies entire categories of projects based on technology. Similarly, by providing a stipulated benchmark, the modified technology matrix approach drastically reduces the cost and effort required of project developers to quantify the baseline. Economies of scale are realized because the same benchmark values can be applied to many different projects. However, the cost of developing and updating the benchmarks, to host countries and institutions responsible for overseeing the market-based projects, may be high. Furthermore, because the benchmarks do not account for circumstances and conditions specific to each individual project, they will in general provide a less reliable estimate of the project counterfactual than the project-specific approach. However, for some specific types of projects the differences in reliability between the two approaches may be insignificant.

The authors believe that the project-specific approach and the modified technology matrix approach should be viewed as complementary, rather than mutually exclusive. Each has its applications. For example, in the case of projects involving the construction of new power plants utilizing advanced, non-commercial technologies, the project-specific approach will tend to reduce to the modified technology matrix approach; i.e., the baseline for such projects will typically prove to be conventional-technology power plants, and some benchmark representing average or typical emissions from these plants will have to be used as the baseline. For these types of projects, the existence of a list of pre-qualified

> **Flexible Protocols**
>
> Exclusive reliance on either the project-specific approach or the modified technology matrix approach could result in lost opportunities. Therefore, a flexible protocol incorporating both approaches has been devised for application to the three case studies.

technologies and stipulated baselines may greatly reduce the costs associated with baseline estimation, without significantly reducing the reliability of the resulting estimate. On the other hand, because of its exclusive focus on new, non-commercial technologies, the modified technology matrix approach may automatically disqualify many worthwhile, additional projects involving conventional technologies. The hypothetical Indian Power Plant Efficiency Improvement project analyzed in Chapter 4, and based on an actual project involving USAID, NETL and other U.S. sponsors, is an example of just such a project. This project would be disqualified under the modified technology matrix approach because it involves the application of conventional technologies to existing facilities; however, a strong argument for its additionality can be developed under the project-specific approach. Also, owing to the costs involved in developing the stipulated baselines, the modified technology matrix approach may not prove feasible or cost-effective for small countries that will host a limited number of market-based projects. However, to overcome the cost barrier, smaller countries could obtain financial assistance from relevant international and bi-lateral sources and/or group together to develop a technology matrix for a select region (i.e. West Africa and Central America) or representing a particular type of countries (i.e. small low-lying island nations).

In short, neither the project-specific approach nor the modified technology matrix approach is a panacea. Exclusive reliance upon one or the other approach may result in significant lost opportunities, as a result either of the expense of implementing the project-specific approach or the automatic disqualification of all projects involving conventional technologies under the modified technology matrix approach. However, a flexible protocol incorporating both approaches should enable the application of the optimal approach in each specific situation. To test this theory, just such a flexible approach is applied to the three case studies presented in Chapters 4 through 6. Our case study analysis indeed lends support to the concept of a flexible protocol. While it was necessary to use the project-specific approach to analyze the Indian Power Plant Efficiency Improvement project (Chapter 4), the application of that approach to the China IGCC project (Chapter 5) or the Argentina Fuel Cell project (Chapter 6) would have proven much more costly than the modified technology matrix approach utilized, without necessarily adding to the reliability of the baseline estimate.

3.2 Generic Baseline Development Guidelines

Having decided upon a flexible protocol for case study analysis purposes, guidelines are required to ensure appropriate selection between the project-specific and modified technology matrix approaches for our three hypothetical projects. This section has three objectives. First, we will present the basic criteria that should determine which of the two approaches to utilize for the three case studies. Because all three projects are electricity supply projects, we will limit our considerations to projects involving electricity generation; however, the approach selection criteria will be defined broadly enough to cover not only the three case studies, but all such power supply projects. Then, we will briefly consider the criteria for pre-qualifying technologies under the modified technology matrix approach, along with procedures for estimating the benchmark. Finally, we will conclude with a few comments concerning the development of protocols for the project-specific approach.

3.2.1 Approach Selection Criteria

Choosing between the two approaches should, for most projects, be a straightforward and fairly obvious process. Table 3.1 presents the various types of projects likely to be implemented in the power generation sector, and the corresponding baseline development approach for each type. Allowable exceptions to the stipulated default approaches are outlined in the third column of the table.

As the table indicates, the modified technology matrix should be the default procedure for analyzing all projects involving the installation of new generating capacity utilizing one of the qualifying technologies. For such projects, there is no need to establish additionality on a project-by-project basis; the application of stringent criteria in the technology qualification process will ensure the additionality of almost all, if not all, projects utilizing qualifying technologies. Furthermore, as argued above, the use of some sort of benchmark as the baseline emissions estimate will be required for most, if not all, projects involving the opening of new capacity. By providing a stipulated benchmark, the modified technology matrix approach will greatly reduce baseline estimation costs for the project developers, while at the same time introducing a degree of standardization and objectivity into the estimation procedure.

As noted in Table 3.1, there are some exceptions to the use of the modified technology matrix approach for evaluating new capacity projects involving qualifying technologies. *First, the approach cannot be used in host countries for which a list of qualifying technologies, or an appropriate set of benchmarks, has not been developed. Small countries, or countries expecting limited market mechanism activity, are likely to fall into this category. Second, project developers should be allowed to utilize the project-specific approach to develop their own baseline, if they so desire, and if they can demonstrate, to the satisfaction of review board, that the baseline thus estimated is more accurate, for their particular project, than the sectoral benchmark.* In such cases project additionality need not be demonstrated using the project-specific approach; rather the project developers may use the modified technology matrix to demonstrate the additionality of their project and the project-specific approach to estimate their baseline.

Finally, if the new capacity is being opened primarily to replace existing capacity or generation, rather than to meet new demand, *and* if it is possible to readily identify the existing capacity or generation being replaced, then the emissions from this existing capacity/generation should be used as the baseline rather than a sector benchmark. For example, suppose that a utility is replacing an old conventional coal-fired plant with a new IGCC plant. In such a case, it should be relatively easy to develop a baseline based on the emissions of the old plant. Furthermore, such a baseline will be more accurate than a sectoral benchmark. Again, in such cases the modified technology matrix can still be used to establish project additionality; only the baseline estimate need be developed using the project-specific approach. The key consideration, in identifying these types of projects, is whether

Table 3.1. Criteria for Selecting an Approach to Baseline Development, for the Electricity Generation Sector

Project Type	Corresponding Approach	Exceptions
1. Projects involving the installation of new capacity, and utilizing advanced qualifying technologies	Modified Technology Matrix	a. Projects to be implemented in host countries without qualifying technology lists/sector benchmarks must use the project-specific approach b. Project developers *may* choose to use the project-specific approach to estimate the baseline, *if* they can demonstrate that the result is more accurate c. Projects designed to replace existing capacity rather than meet new demand should use the project-specific approach for baseline development, *if* the capacity to be replaced can be readily identified.
2. All projects utilizing conventional, non-qualifying technology	Project-specific	a. Projects involving the installation of new capacity to meet new demand should use a sectoral benchmark for baseline estimation, *unless* the project-developers choose to use the project-specific approach *and* can demonstrate that the result is more accurate.
3. Projects involving the retrofitting of advanced qualifying technologies to existing facilities, with no resulting change in capacity	Modified technology matrix to establish additionality; project-specific to estimate the baseline	None.

or not the new capacity is being built to meet new demand, or to replace existing capacity; if the former, the modified technology matrix is the default baseline estimation procedure; if the latter, and if the capacity to be replaced can be clearly and readily identified, the project-specific approach should be used.

For all projects involving non-qualifying technologies, the project-specific approach must be utilized. These projects should not be rejected out of hand; many projects utilizing conventional technologies may be additional, due to the specific circumstances and barriers surrounding these projects within the respective host countries. In many cases, such projects will provide the most cost-effective means of reducing emissions, precisely because they rely on existing technologies that are well understood. Therefore, project developers should be given the opportunity to demonstrate their additionality, using what will be the only means at their disposal in these cases: the project-specific approach.

As far as determining additionality is concerned, there are no exceptions for projects utilizing conventional technologies: the developers of such projects *must* employ the project-specific approach. However, certain types of conventional-technology projects should utilize the sectoral benchmarks developed under the modified technology matrix approach as their baseline estimates. Specifically, projects involving the installation of new generating capacity, designed primarily to meet new demand, should employ the modified technology matrix benchmarks, unless the project developers choose to utilize the project-specific approach, and can demonstrate that the resulting baseline estimate is more accurate.

Finally, there is a third class of projects not considered above: those involving the retrofitting of advanced, qualifying technologies to existing power plants without either adding new capacity or replacing existing capacity. The additionality of these projects may be established based on the modified technology matrix. However, the project-specific approach should be used to develop the baseline rather than a sectoral benchmark, since the emissions of the power plant prior to the retrofit are likely to provide the best basis for the baseline. In practice, such projects are likely to be very few in number, if not non-existent.

3.2.2 Development of the Technology Matrix

The application of the modified technology matrix by project developers will require little in the way of guidelines. All the developers need do is provide evidence that their project utilizes a qualifying technology, and then select the stipulated benchmark appropriate to their particular sector, subsector, and type of project.

However, development of the modified technology matrix, by those responsible for implementing market mechanisms, will be a difficult and complex undertaking. Chapters 5 and 6 present two examples of technology matrix development, for two specific technologies (IGCC and fuel cells) in two different host countries (China and Argentina). Because development issues are considered in detail in these two examples, we will limit our discussion at this point to a consideration of a few general principles that were used to guide the case study analysis

First, in order to limit errors as much as possible to those types that may be inherent to the modified technology matrix approach, stringent additionality criteria were applied to qualifying the technologies for the two case studies. If error minimization is the goal, it is necessary to identify and include only those technologies that will, in the vast majority of cases, prove non-commercial in the absence of market mechanism incentives. If certain technologies that are additional in general, but non-additional in many particular instances, make it onto the list, then it is likely that developed country investors will preferentially fund the qualifying non-additional projects. (These projects will, almost by definition, be more attractive than the additional projects from an economic standpoint). It is true that the application of stringent criteria may exclude a number of technologies that will prove additional more often than not. Keep in mind, however, that under the guidelines presented in Table 3.1, project developers will still be given the opportunity to qualify these projects on a case-by-case basis, using the project-specific approach. Because it is a generic approach that qualifies entire classes of projects, it will be impossible to prevent all non-additional projects from qualifying under the modified technology matrix. However, to the extent that error reduction is the goal, care must be taken to ensure that the number of projects qualified erroneously is kept to a minimum.

Given this same goal, the modified technology matrix should be developed on a country-by-country basis. Developing countries are distinguished by large differences in investment priorities, financial markets, fuel and resource availability, electricity tariffs, regulatory climates, technological sophistication, cultural responses to risk, etc. Given these differences, a technology that may prove additional in the vast majority of countries will not necessarily be additional in *all* countries. Furthermore, when comparing the economics of a candidate technology against comparable commercial technologies at a generic level, the estimated differences in such metrics as costs, payback period, and internal rate of return should be sufficiently large to ensure that, with few exceptions, these differences will not be offset by particular circumstances. Similarly, when considering a technology from the standpoint of market penetration, the criteria for inclusion should be stringent: evidence of more than a few commercial-scale projects utilizing a particular technology without government funding throughout the world should be sufficient to exclude a particular technology from the matrix. Ideally, a technology should be qualified only if meets both the economic comparison and market penetration tests.

The stipulated baselines should be developed at a highly disaggregated level, to ensure that an applicable, comparable baseline will in fact be available for each of the myriad projects likely to qualify under the modified technology matrix approach. In general, the stipulated baseline for each qualifying technology should represent, as closely as possible, the likely alternative to that technology in the particular host country. Because the modified technology matrix will be applied exclusively to projects involving the commissioning of new capacity to meet new demand, the stipulated benchmark should correspond to newer power plants utilizing commercial technology. For example, the benchmark for an IGCC project in China might be developed based on the average emissions of conventional coal-fired power plants commissioned in China in the last five years.

These general principles for benchmark stipulation and technology qualification will be developed in more detail in Chapters 5 and 6, where the modified technology matrix approach will be applied to two illustrative examples. It should be emphasized that these general principles apply only if error minimization is the goal (as is the case in our case study analysis). In order to strike a balance between cost and accuracy it may prove necessary to relax these principles if and when the modified technology matrix is applied to actual market-based projects.

3.2.3 Some General Comments Concerning Standardization of the Project-Specific Approach

The project-specific approach is not readily amenable to the development of detailed protocols or standardized formulas. Attempts to apply detailed, standardized procedures may in some sense violate the basic idea underlying the approach: as its name implies, the project-specific approach involves the tailoring of the baseline estimate to the specific project under consideration. The project-specific approach is based on the recognition that each project is unique, and that the most reliable estimate of the project counterfactual can be developed only through a careful, detailed analysis of all the site-specific factors that determine the project's uniqueness. Such an analysis may lead to a final conclusion far removed from the *a priori* expectations inherent in a standardized procedure or set of algorithms. The Indian Power Plant Efficiency project, which is analyzed using the project-specific approach in Chapter 4, provides an excellent example. When we began our analysis of this project, involving efficiency improvements at India's coal-fired power plants, we fully expected the project counterfactual to be the same power plants prior to improvement. As we shall see, our analysis of circumstances specific to the Indian power sector led us ultimately to a much different counterfactual.

Thus any attempt to apply detailed standards in this context runs the considerable risk of defeating the purpose of the project-specific approach. It would be extremely difficult, if not impossible, to *imagine* in advance all of the project-specific nuances that should be considered when developing the baseline. If protocols are to be attempted for the project-specific approach, this attempt should probably be made in the concrete rather than the abstract. Standardized protocols should be developed from the bottom up, on the basis of close analysis of a large number, and wide variety, of real and/or hypothetical projects. It is entirely possible that such an exercise would lead to the conclusion that protocols should be limited to nothing more than general, qualitative procedural guidelines.

Thus, our analysis of the Indian Power Plant Efficiency Improvement project in the following chapter represents at best a first step towards the ultimate goal of standardized protocols; it is *not* designed to provide an illustrative example of the application of such protocols. Nonetheless, we did develop some rough, non-comprehensive, qualitative procedural guidelines during the course of our analysis of this project. These guidelines will be presented in the next chapter, within the context of our analysis.

3.3 Summary

In this chapter, we presented our flexible approach to baseline estimation for the three case studies, incorporating both the project-specific and modified technology matrix methods. We also presented some basic guidelines to aid project developers in selecting a particular methodology for a given project type (Table 3.1), and we presented some general principles for technology matrix development. Finally, the potential difficulties and pitfalls associated with protocol development for the project-specific approach were addressed.

In the following three chapters, the guidelines developed in this chapter are applied to the three example projects: a power plant efficiency improvement project in India (Chapter 4), an IGCC project in China (Chapter 5), and a fuel cell project in Argentina (Chapter 6). As explained in Chapter 1, our basic approach to the three projects involves two steps. First (Chapters 4 through 6), we adopt the subjective viewpoint of the project (or technology matrix) developers, and attempt to construct additionality arguments and baseline estimates in as rigorous a manner as possible. Second (Chapter 7), we submit the three case studies to an objective critique, including a qualitative error assessment. Because we wish to limit the error assessment, to the extent possible, to those types of errors that may be inherent to the baseline development problem and the two generic approaches to that problem, it is necessary to proceed with the baseline estimations in as rigorous a manner as possible. Only by approaching baseline development for the three projects with some rigor can we ensure that the error assessment excludes errors resulting from a mere lack of diligence.

4. EMISSIONS BASELINE DEVELOPMENT FOR THE INDIAN POWER PLANT EFFICIENCY IMPROVEMENT PROJECT

4.1 Introduction

The Indian Power Plant Efficiency Improvement project represents a hypothetical replication of an ongoing project--the Greenhouse Gas Pollution Prevention Project (GEP). The GEP project has a number of objectives beyond power plant efficiency improvement, including plant availability improvement, electrostatic precipitator performance improvement, assessment of the impact of coal quality on plant performance, and commercialization of large scale coal ash utilization technologies. The efficiency improvement activities are being conducted under the Efficient Coal Conversion (ECC) component of the GEP project. The primary sponsor of the GEP project is the U.S. Agency for International Development (USAID); in addition, the project team includes the NETL, the Electric Power Research Institute (EPRI), the Tennessee Valley Authority (TVA), and India's National Thermal Power Corporation (NTPC).

In developing our hypothetical project, we have used the GEP project as a starting point. Specifically, we hypothesize that a number of U.S. companies have formed a joint venture to replicate the GEP project at power plants operated by India's State Electricity Boards (SEBs). This hypothetical replication of the GEP project by private investors adheres closely to the project activities comprising the GEP project; however, some elements of the hypothetical project have been changed to place it in the context of an operational market-based activity regime and to illustrate specific baseline development issues.

4.2 Project Description

4.2.1 The GEP Project

Before providing a description of the hypothetical case study, we will begin with a factual account of the project upon which it is based – the GEP project. The goal of the GEP project is to reduce carbon dioxide emissions by improving the efficiency of existing coal-fired power plants. The U.S. partners in the project, including USAID, the Department of Energy (DOE), NETL, EPRI, and TVA, among others, are providing a portion of the required funding, equipment, and technical training. Specifically, USAID is providing approximately $4 million in funding, NETL is providing labor and other resources at no cost, and EPRI and TVA are providing various documents at no cost. It should be noted that the NTPC has joined EPRI's power plant performance business unit, and is receiving heat rate improvement information through this association.

The project involves systematic performance monitoring and diagnostic testing of the boilers, turbines, condensers, and auxiliary equipment at NTPC coal-fired power plants. The tests enable identification of specific plant components that are operating at less than design or optimal efficiency. The tests also provide some indication of the corrective action required to improve component performance. The needed corrective actions will sometimes require capital expenditures, e.g., to replace worn equipment or to upgrade equipment; however, in many cases the required actions will involve procedural changes in plant operation rather than capital improvements. For the most part, the project relies upon existing commercial technology, although in a few instances advanced technologies may be utilized.

The GEP project is by no means the first effort to improve heat rates at Indian power plants. The NTPC, and even some of India's State Electricity Boards (SEBs), had begun heat rate improvement efforts prior to the GEP project. However, the GEP project differs from these earlier efforts in that it is a *systemic* approach to monitoring and diagnosing efficiency losses within each component of the power plant. The performance monitoring equipment and test procedures are being introduced into India for the first time through the project. Training of NTPC personnel constitutes a major component of the project to ensure that the monitoring and testing program continues after the U.S. engineering teams have completed their in-country work. Dissemination of the knowledge and information gained through the project is another major component.

The project was initiated at NTPC's Dadri power plant, which consists of four coal-fired units, each with a capacity of 210 MW. The Dadri plant is one of the newer and most efficient plants in India. In fact, two of the four units began operation in the past five to six years, and the plant's overall efficiency prior to project initiation was approximately 33 percent – comparable to the U.S. average for coal-fired plants. The objective in choosing Dadri for the initial efforts is to demonstrate that the project approach will yield efficiency gains at *all* power plants – even plants that are already operating at high efficiency.

A long-term goal is to extend the project beyond NTPC to power plants owned and operated by India's State Electricity Boards (SEBs). The project is directly replicable at over 130 units similar to NTPC's 200/210 megawatt (MW) coal-fired units, and, with minor modifications, the project approach could be applied to nearly 60,000 MW of thermal capacity. If extended to all of India's thermal power plants, the project approach could improve India's average heat rate by an estimated 4.5 percent. Ultimately, the approach could also be applied to power plants in other developing and developed countries.

The work at Dadri, which is now completed, resulted in an overall efficiency improvement of 1.8 percent. Because of the success at Dadri, the project has been expanded to cover the following NTPC plants (all coal-fired):

- Rihand (1000 MW)

- Singrauli (2000 MW)

- Badarpur (705 MW)

- Wanakbori (1260 MW)

66

- Kahalgaon (840 MW)--mill performance optimization only

- Vindyachal (1260 MW)

- Ramagundam (2100 MW).

In addition to these NTPC plants, the project team has received requests from SEBs and Independent Power Producers (IPP) to replicate the project at a number of other plants. The ultimate goal is to replicate the project at all of India's thermal power plants.

Some of the specific activities performed under the project include the following:

- Use of power plant modeling software for performance optimization

- Accurate measurement of unburnt carbon in fly ash using iso-kinetic and high-volume samplers (based on iso-kinetic sampling technology)

- Use of water-cooled high velocity thermocouple probes, also known as suction pyrometers, for combustion characterization

- Performance optimization using a portable Data Acquisition System (DAS) and on-line computational software to conduct tests of high-pressure heaters, low-pressure heaters, condensers, high-pressure/intermediate pressure efficiency, enthalpy drop, boiler feed pumps, and boilers

- Condenser air-in-leakage testing with helium leak detector

- Validation of condenser back pressure

- Demonstration of U.S. on-line condenser cleaning system.

Unlike many other heat rate improvement projects, the India project is designed to maintain heat rates at lower levels once these levels have been achieved. This is accomplished by providing the monitoring and diagnostic equipment, and the required training, needed to perform periodic performance evaluations. In particular, emphasis will be placed on periodic condenser air-in-leakage detection and elimination, periodic condenser cleaning, and the continued use of mathematical modeling software for optimization. The project team has identified combustion optimization and condenser and coal water (CW) system optimization as the two areas with the greatest potential for efficiency improvement at India's thermal power plants.

A centralized organization – the Centre for Power Efficiency and Environmental Protection (CENPEEP) – has been created within the NTPC to ensure that the testing program will be maintained even after U.S. involvement in the project has ended. CENPEEP is the NTPC institution responsible for implementing the project. It is a permanent organization within the NTPC, with a 100 percent dedicated staff.

It should be emphasized that the GEP project itself includes additional activities beyond the heat rate improvements described above. In addition to the Efficiency Task, the GEP project includes an Environmental Improvement Task, aimed primarily at electrostatic precipitator performance; a Monitoring and Diagnostics Task, aimed at improving power plant availability; a task to study and assess the impact of coal quality on plant performance, operation, and maintenance; and a task aimed at the commercialization of large-scale coal ash utilization technologies. Moreover, in addition to these power sector activities, another component of the GEP project is aimed at encouraging the use of bagasse and other biomass fuels in cogeneration applications in the sugar industry. However, so that we may focus specifically on the area of greenhouse gas emission reductions in the power sector, our hypothetical case study, described below, has been limited to those activities designed specifically to improve power plant heat rates.

4.2.2 The Hypothetical Indian Power Plant Efficiency Improvement Project

For purposes of illustrating the baseline estimation process for a heat rate improvement project in India, the GEP project has been selected as the basis for the development of a hypothetical case study which will be referred to as the Indian Power Plant Efficiency Improvement project. This case study supposes that a number of U.S. electric utilities have formed a joint venture to replicate the GEP project at SEB plants located throughout India. The U.S. utilities will receive all credits to be awarded under the market-based activity for their financial participation in the project; the SEBs will reap the direct benefits of the project – i.e., the benefits arising from the efficiency improvements. The U.S. utilities will provide the majority of the financing for the project, while the SEB contribution will be limited mainly to time and labor.

The project activities will replicate the heat rate improvement activities undertaken as part of the GEP project. Again, these activities involve systematic performance monitoring and diagnostic testing of the boilers, turbines, condensers, and auxiliary equipment at the subject power plants. Based on these tests, the corrective actions needed to return the plants to their design efficiencies will be identified and undertaken. As in the case of the GEP project, the Indian Power Plant Efficiency Improvement project will rely primarily upon existing commercial technologies. Training of host country personnel, to ensure that the efficiency gains are maintained, will constitute a major component of the project.

4.3 Emission Baseline Development

4.3.1 Evaluation of Baseline Options

The first step in developing an emission reduction estimate for the project is to select a baseline development approach. As discussed in Chapter 3, we are, for case study purposes, limiting our consideration to two generic approaches for baseline development--the modified technology matrix approach and the project-specific approach. The selection of one or the other of these approaches is dependent on the type of project involved. Table 3.1 in Chapter 3 presents the recommended selection criteria for various project types. Reviewing this table, it should first be evident that the Indian Power Plant Efficiency Improvement project falls into the second category of projects listed in the left-hand column; i.e., it involves conventional technology that will not automatically qualify as additional under the modified technology matrix approach. In fact, much of the plant improvement is to be achieved by changes in plant operating procedures rather than equipment replacement or modification.

As Table 3.1 indicates, the recommended baseline development approach for Type 2 projects is the project-specific approach. The exception noted in the table applies only to projects involving the installation of new capacity, and hence it is not relevant in our particular case. Leaving out the noted exception, Type 2 projects in general, and the Indian Power Plant Efficiency Improvement project in particular, are not amenable to the modified technology matrix approach, for two reasons. First, because the project involves improvements to an *existing* power plant (other than a plant life extension), the best baseline for the project will likely be an estimate of the plant's emissions without the improvements. As suggested in Chapter 3, the stipulated baselines utilized in the modified technology matrix approach are better suited to projects involving the opening of new power plants than to projects involving modifications to existing facilities. For the latter projects, project-specific baselines based on the existing facilities' emission rates can be developed at a low cost, and such baselines will generally prove more accurate than the standard, stipulated baselines used in the modified technology matrix approach.

Second, and more importantly, the additionality of the project cannot be demonstrated based on a consideration of technology alone. To qualify as additional under the modified technology matrix approach, a project must utilize advanced non-commercial technology that is not economically viable without government subsidies or other favorable financing terms or has not reached a particular market penetration rate. But the power plants included in the project are conventional coal-fired facilities, and for the most part the project involves procedural changes and the utilization of commercial technology to affect plant performance improvements. The additionality of this project therefore can be demonstrated only through the application of the more rigorous project-specific approach. In short, the modified technology matrix approach is not a viable option for this particular project. Therefore, the project-specific approach must be used.

4.3.2 Additionality

Having determined that the project-specific approach will be followed, the next step in the baseline development process is to address the issue of additionality. As discussed in Chapter 3, we are suggesting a flexible approach to testing for additionality under the project-specific option. Under this approach, a project's additionality may be demonstrated either through an economic feasibility analysis, or by providing evidence of barriers (regulatory, technological, cultural, knowledge, etc.) that would prevent the project from being undertaken without the market-based activity. This flexible approach recognizes that a variety of factors, beyond project economics, may prevent projects from being undertaken, and that the market- or project-based activity may enable a project's sponsors to overcome both the economic and other barriers to project implementation. Figure 4.1 provides a flowchart illustrating the procedure used to establish additionality for the Indian Power Plant Efficiency Improvement Project.

The Indian Power Plant Efficiency Improvement project will not qualify as additional based on a consideration of economics alone. As previously noted, the GEP project, upon which the hypothetical project is based, is designed to be replicable throughout the Indian power sector; as such, it must prove economically viable even without the favorable financing terms being provided by the U.S. partners. The heat rate improvements to be gained as a result of the Indian Power Plant Efficiency Improvement project will enable the SEBs to reduce fuel consumption and fuel costs. These cost reductions will be gained in part through procedural changes in plant operations rather than capital expenditures; in fact, the project is designed to identify actions that will yield significant efficiency improvements at relatively low costs. Furthermore, although capital will be required for the performance monitoring and testing equipment and for some of the needed plant improvements, it is expected that the payback period for the capital investment will be short (and would be short even without the U.S. funding). Hence, the project will not qualify as additional on the grounds that it would be subeconomic without the U.S. partners' financial investment.

Nonetheless, other barriers exist which may have prevented the project from being replicated were it not for provisions of the market-based activity. Two such barriers are of particular importance: the financing barrier and the knowledge barrier. Our approach, in the following pages, will be to attempt to establish the additionality of the project on the basis of these two barriers. We will begin by considering the financial barrier, and then proceed to a consideration of the knowledge barrier. Finally, we will address the temporal issue as it relates to the additionality of the project.

4.3.2.1 The Financial Barrier. Throughout the United States and the world, numerous opportunities to improve the efficiency of energy production, transformation, transmission, and utilization remain unfulfilled. This is true despite the fact that many such opportunities would yield savings well in excess of the costs required for their implementation. Clearly economic feasibility is not the barrier preventing such projects from being undertaken. What, then, are the barriers to such projects? In many cases, opportunities to improve efficiency may simply go unrecognized. Failure to recognize opportunities for efficiency improvement is a particularly important barrier in the energy end-use sector, where technical knowledge concerning energy is often limited (especially among residential

70

users). However, knowledge tends to be a less common (though by no means rare) barrier in the energy production and electricity generation sectors. In these sectors, the unavailability of capital is often the key barrier preventing otherwise economically viable projects from being undertaken. This is especially true in developing countries, where opportunities abound but investment capital is limited, and energy efficiency is often a low investment priority relative to more urgent needs. This describes the situation in India.

Within the electricity sector, India's urgent need is to increase its capacity to meet existing and projected demand. Currently, India faces acute electricity shortages. Power outages average 20 percent of demand during peak use hours and 10 percent of off-peak demand. Furthermore, India's electricity demand is expected to grow by seven percent per year through 2005, and 4.4 percent per year thereafter (through 2020). To alleviate the electricity shortages and meet new demand, the government plans to add 40,000 MW of capacity by 2002. About half of this total is expected to be funded by private – mostly foreign – sources, but this still leaves about 20,000 MW to be funded by the government (either directly, or through the various Federal and State utilities). Furthermore, the government estimates that an additional 71,500 MW of capacity will be required between 2002 and 2007. To put these projections in perspective, India's total capacity is only 85,000 MW at present. Clearly, most of the capital available to the electricity sector must be utilized to build the capacity needed to meet India's large and growing electricity demand.

Still, efficiency improvements at existing power plants could be a low-cost, effective means of increasing generation. In developed countries, where electricity supply and demand are in balance, heat rate improvements generally result in a reduction in the amount of a fuel consumed to meet a given load. But in developing countries such as India, where a significant supply shortfall exists, heat rate improvements may actually lead to an increase in generation for a constant level of fuel supply. In fact, increasing generation through power plant heat rate improvements may represent a low cost alternative to the building of new capacity.

Could the SEBs obtain the financing needed to increase generation through efficiency improvement without the aid of the U.S. project partners? The prospect of raising the required capital is very poor. As a result of a tariff structure that keeps tariffs artificially low for residential and agricultural users, heavy debts, poor revenue collection, and inadequate investment, the financial viability of most of the SEBs is already poor. As of 1997/98, the SEBs overall rate of return stood at -17.6 percent. Furthermore, the combined commercial losses of the SEBs reached 2.3 billion dollars in 1996/97. These losses were actually less than half of what they would have been, had the SEBs paid their dues to other government utilities and enterprises. Three of the country's nineteen SEBs (Delhi, Bihar, and Uttar Pradesh) together owe 3.4 billion dollars to the central power utilities (including the NTPC), and the remaining 13 SEBs owe an additional 3.4 billion dollars.

Figure 4.1. Determining Additionality of the Indian Power Plant Efficiency Improvement Project

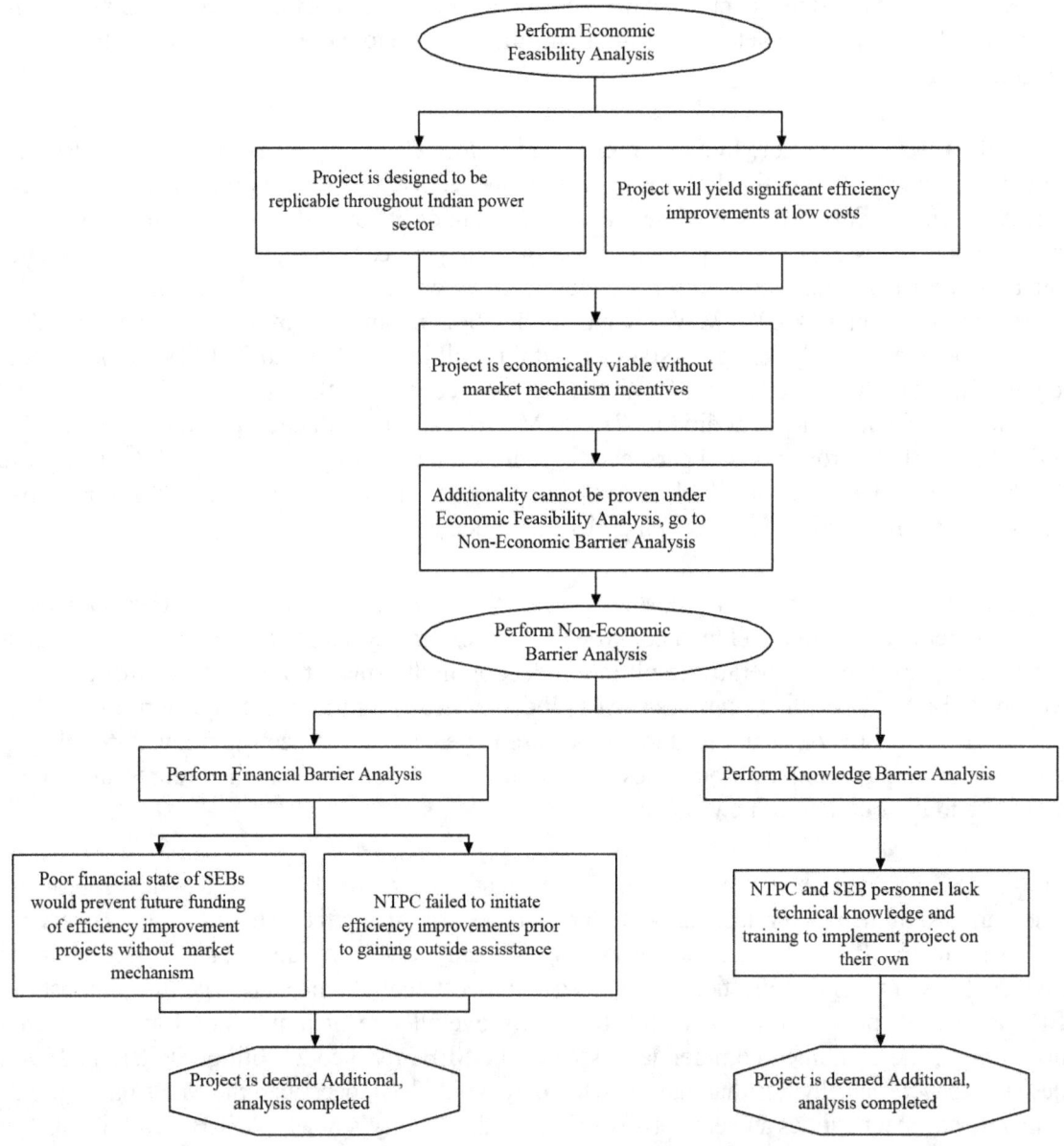

Subsidies represent a large part of the problem. Subsidies were originally introduced to promote the economic development of some sectors (especially the agricultural sector). These early subsidies were small and could be funded out of the government's budget. During the 1980s and 1990s, subsidies increased many fold and could no longer be funded by the government, but instead appeared as losses to the SEBs. The "effective subsidy" to the agricultural and residential sectors (i.e., the difference between costs and revenues earned), was assessed at nearly 1.4 percent of India's entire GDP in 1996/97. Low tariffs are also contributing significantly to the SEBs poor financial health. Current tariff levels are estimated to cover only 80 percent of the total cost of supply.

The heavy losses experienced by the SEBs in recent years would be much larger were it not for bailouts being provided by the State governments. Even those few SEBs that have managed to show a profit have done so only because of the bailouts received from the States. For example, the Maharashtra SEB showed a profit of $45 million in 1996/97, but received government subsidies of $61.6 million during the same period.

The SEBs financial problems are reflected in their poor operating performance. The average plant load factor for SEB power plants, which represent approximately 75 percent of national generating capacity, is only 58 percent, as compared with the national average of 63 percent. Transmission and distribution losses have been estimated as high as 23 percent.[22]

According to the Indian government's Ninth (and most recent) Five-Year Plan, without power sector restructuring and tariff rationalization, the internal resources generation of the SEBs will be negative 18 billion dollars over the next decade. As the Ninth Five-Year Plan notes, this raises serious questions as to whether the SEBs will be able to contribute their share of capacity additions as envisaged in the Plan and thereafter. The poor financial health of the SEBs has already been acting as a significant constraint on power sector investment. The Eighth Five-Year Plan called for an additional 2,810 MW to be financed by the private sector; however, only half of this goal (1,430 MW) was met. In assessing the reasons for this shortfall, the Ninth Plan states:[23]

> "The shortfall in the private sector was due to the emergence of a number of constraints which were not anticipated at the time the policy was formulated. The most important is that lenders are not willing to finance large independent power projects, selling power to a monopoly buyer such as SEB, which is not financially sound because of the payment risk involved if SEBs do not pay for electricity generated by the IPP.
>
> Some small projects were able to get financing because the payment risk was deemed

[22]Tata Energy Research Institute, "Indian Power Sector: Change of Gear," http://www.teriin,org/energy/power.htm.

[23]Government of India, Ninth Five-Year Plan.

acceptable. Five projects have received Central Government counter-guarantees and two more are eligible subject to resolution of other problems. The Central Government will not provide counter guarantees for power projects in the future. A number of private sector projects are seeking credit enhancement by escrowing certain receipts of the SEBs to assure payment of IPP dues. The scope for escrow arrangements is limited and in any case they do not increase the overall financial viability of the SEB. Rather, by earmarking part of the SEB's existing receipts for new private sector projects, they reduce the viability of the rest of the system.

The ability to attract private investment into the power sector on a significant scale in the future therefore depends crucially upon bringing about improvements in the financial condition of SEBs."

According to the Tata Energy Research Institute (TERI), "the unsatisfactory financial health of the SEBs has acted as a constraint to provide adequate investments for improving the utilization of existing capacities and for new capacity creation."[24]

In conclusion, it appears highly unlikely that the SEBs would be able to attract the investment moneys needed to fund the Power Plant Efficiency Improvement project, absent the market-based activity. Thus we conclude that the emission reductions to be achieved at the SEB power plants, as a result of the project, qualify as additional.

4.3.2.2 The Knowledge Barrier. [25] The lack of potential capital investment sources beyond the U.S. partners in the project, although sufficient in and of itself to establish additionality, is not the only barrier that the project would face in the absence of the market-based activity. In addition, SEB personnel lack the technical knowledge and training necessary to implement the project on their own. In establishing this lack of knowledge as a real barrier to the project, it is necessary to describe and consider carefully the state of training and understanding of the testing, monitoring, and diagnostic procedures within the NTPC prior to implementation of the GEP project. Evidence of a lack of knowledge of these procedures within the NTPC will provide a strong indication of a similar lack of knowledge within the SEBS, since the latter utilities are in general less sophisticated than the NTPC.

Knowledge of the GEP plant testing equipment and procedures was not wholly lacking within the NTPC prior to project initiation. NTPC personnel were aware of at least some of the testing procedures now being utilized as part of the GEP project. In fact, some NTPC power plants had performed some of the tests on their own.

---

[24]Tata Energy Research Institute, "Indian Power Sector: Change of Gear," http://www.teriin.org/energy/power.htm.

[25]The information on the state of training at the NTPC used in this sub-section was obtained via personal communication with Scott Smouse of NETL.

However, knowledge of the testing procedures was limited to only a few NTPC personnel, scattered throughout the organization. And again, knowledge was limited to only a few specific tests; other tests that are being implemented as part of the project were new to the NTPC and to India. But most importantly, NTPC personnel had never combined the individual tests into an integrated, systematic approach to improving heat rates. Rather, individual tests had been performed on an ad hoc basis, usually in response to a specific perceived problem. For example, if the turbine was believed to be performing poorly at a particular power plant, tests would be conducted on the turbine to verify and diagnose the problem, but testing of other plant equipment (e.g., the boiler and mills) would not be performed. Furthermore, testing had not been conducted uniformly across all of the NTPC's plants; limited testing had been performed only at individual power plants with staff knowledgeable in the specific procedures to be performed. There was no central, organization-wide commitment to testing of the type being performed under the GEP project, nor was there a centralized staff dedicated to performing the tests in a systematic manner across all NTPC power plants. In fact, a key objective of the GEP project was to create a central organization within the NTPC that, once trained, would eventually assume sole responsibility for conducting and maintaining the testing program; that objective has been fulfilled by the creation of CENPEEP.

In short, NTPC staff possessed limited knowledge of some components of the testing program, but the idea of combining the tests into an integrated, systematic testing program, to be applied throughout the NTPC by a dedicated, centralized organization, was new. The idea of an integrated testing program, and the knowledge and training necessary to implement such a program, was brought to the NTPC by the U.S. partners in the project. TVA, in particular, had used this approach at its own power plants for some time, as have many other U.S. and foreign utilities. It is, of course, possible that the NTPC would *eventually* have conceived such a program on its own, or in conjunction with other U.S. or non-U.S. partners. However, the fact remains that, up until the initiation of the GEP project, the NTPC did *not* conceive of such a testing program. Furthermore, even if the idea for such a program had existed, implementing the idea would have required the acquisition of knowledge and skills not available within the NTPC's in-house staff. Recall, again, that although NTPC personnel had knowledge of some of the test procedures and equipment that are being used in the project, other tests are new to the staff. In particular, many of the data acquisition tools that are being used as part of the project had never been used in India before, and the data interpretation skills that have been required during the project were, prior to project implementation, lacking. Furthermore, even in the case of test equipment and procedures that had been used at some NTPC power plants, significant additional training was required to expand the knowledge base beyond the small number of NTPC personnel familiar with the equipment/procedures to allow country-wide adoption. As noted previously, training of in-house staff represents a major objective of the project.

The training requirements for implementation of the project are quite substantial, as evidenced by the amount of training that has actually been conducted under the ongoing GEP project. As of early 1999, 12 technical teams from the United States have provided over 3,200 labor hours of training and technical assistance. As a result of this effort, Indian power plant personnel have acquired over 28,000 hours of training. Furthermore, training is still ongoing, and will continue into the planned follow-on Phase II of the GEP project.

We conclude that the NTPC lacked the knowledge and skills necessary to implement the GEP project without the aid of the project's U.S. backers. It is possible that this knowledge barrier might have been circumvented at some future point in time; we will return to this possibility in the next subsection.

Thus far, we have focused our attention on the knowledge barrier as it existed at the National Thermal Power Corp. Prior to the implementation of the GEP project. In the case of the State Electricity Boards (SEBs) serving as hosts to the Indian Power Plant Efficiency Improvement project, the existence of the knowledge barrier is more evident and easier to demonstrate. In general, personnel at India's SEBs are not as well trained as NTPC personnel. For example, whereas NTPC personnel had knowledge of, and had even performed, some of the tests that are being utilized in the GEP project, the SEBs have rarely used these tests in the past, and lack the knowledge and training required to perform the tests. It is conceivable that the SEBs could obtain the required training from sources other than the project's U.S. partners, but this would require the expenditure of funds, and, as discussed previously, the financial barrier represents a formidable obstacle for the SEBs.

In conclusion, the technical knowledge required to implement the project without the aid of the U.S. partners is lacking within the SEBs. It may seem trite but it is nonetheless true that neither an individual nor an organization can implement a project that they do not know how to implement. The knowledge barrier is a real and formidable barrier. Its existence at the SEBs would prevent the project from being undertaken were it not for the aid of the U.S. partners, to be provided in return for the emission reduction credits that will be awarded under the market-based activity.

4.3.2.3 Temporal Considerations. Although the financial and knowledge barriers prevent the project from being undertaken without the market-based activity *at present*, these barriers will not necessarily continue to exist indefinitely. The passage of India's Electricity Regulatory Act of 1998, and the consequent creation of the Central Electricity Regulatory Authority, may prove to be the beginning of a successful attempt to restructure India's power sector and improve its financial health. If so, it is possible to imagine a future in which the gap between electricity supply and demand has been closed and improving the efficiency of existing power plants

> **Key Point**
>
> At some point in the future, the knowledge and financial barriers may be eliminated. Therefore, the issue of additionality for the India Power Plant Efficiency Improvement Project should be re-evaluated on a periodic basis.

has become an investment priority. The future elimination of the knowledge barrier seems even more likely. Given that NTPC personnel already possessed some knowledge of the tests being performed under the GEP project, it seems probable that further diffusion of knowledge within the organization would eventually have provided the utility with the technical ability to develop and undertake a similar project on its own. Knowledge diffusion would likely have been a slower process in the case of the SEBs, but even here, it is possible that the knowledge barrier would eventually have been eliminated even without the help of the U.S. partners.

While it is clearly possible that the existing barriers to the project would eventually have been eliminated, it is impossible to know, in advance, whether or when this might have happened.

Therefore, in the case of this project, it is recommended that the issue of additionality be re-evaluated on a periodic basis. Specifically, the project partners should resubmit the project for qualification as an additional project on a periodic basis (perhaps once every five years). To retain the project's status as additional, the project partners should demonstrate, every five years, that the financial circumstances and investment priorities of the SEBs continue to act as a project barrier. If, at some point, this can no longer be demonstrated, the project would lose its additionality status and would no longer be awarded emission reduction credits.

Ideally, the additionality re-qualification process would involve a reassessment of the knowledge barrier as well as the financial barrier. However, given that the project itself is designed to eliminate the knowledge barrier, it will not be possible to determine whether or not this barrier would have continued to exist but for the project, once the project has begun. Instead, re-qualification must be based exclusively on a consideration of the financial barrier. This approach implicitly assumes that the knowledge barrier would not, in the absence of the project, outlive the financial barrier. Such an assumption seems reasonable, given the severity of the financial problems facing the power sector and the fact that some knowledge of the tests to be performed as part of the project already existed within India (at the NTPC). Arguably, the knowledge barrier would have been eliminated before the financial barrier, in the absence of the project.

To summarize, the additionality of the Indian Power Plant Efficiency project has been established on the basis of two barriers that would have prevented the project from being undertaken in the absence of the market-based activity: a financial barrier and a knowledge barrier. On the basis of this initial additionality assessment, it is proposed that the project receive emission reduction credits for a five-year period beginning at present. After five years, a reassessment will be performed to determine whether or not the financial barrier still exists. If the project partners are able to establish its continued existence, the project will be awarded emission reduction credits for another five years; otherwise, the project will cease to qualify for credits. In like manner, reassessments of the project's additionality status will continue to be made every five years.

4.3.3 Baseline Development

4.3.3.1 Establishing the Qualitative Baseline. Having established that the project meets the additionality criterion, the next step is to establish the project's baseline. Under the project-specific approach, the baseline should represent a projection of what emissions would have been "but for the project." Hence, the first question that must be addressed is a qualitative question: what would have happened had the project not been undertaken? We would like to propose a standard step-by-step procedure for addressing this "what if" question. Although perhaps not suitable for all types of projects, we believe that, for many project types, this standard approach will provide an efficient and reliable means for establishing the most likely alternative to the project. The proposed step-by-step procedure is as follows:

1. Formulate a working hypothesis concerning the most likely alternative to the project.

77

2. Identify, as comprehensively as possible, other project alternatives.

3. Assess the likelihood of the first alternative identified in step 2 vis a vis the hypothesized alternative

4. Based on this assessment, either reject the alternative, or accept it as the new working hypothesis

5. Repeat Steps 3 and 4 until the list of project alternatives identified in Step 2 has been exhausted. The final hypothesized alternative is accepted as the qualitative baseline.

This generic procedure, and its application to the project, is illustrated in Figure 4.2.

Let us now illustrate the procedure by applying it to the Indian Power Plant Efficiency Improvement Project. The first step is to develop a working hypothesis as to the most likely alternative to the project. In the case of a project such as this, involving relatively low-cost modifications to existing facilities, the most likely alternative may seem fairly obvious: the SEBs would have continued to operate the power plants without the modifications. Relative to a project involving, say, the construction of a new power plant designed to meet new demand, the efficiency improvement project was driven by a less compelling need. Rather, the project appears to be voluntary in character; it will yield benefits in excess of its costs, but these benefits could have been forgone without dire consequences.

The hypothesized most likely alternative, therefore, would have been to do nothing. Having developed the initial working hypothesis, the next step is to identify other possible project alternatives. A consideration of the Indian Power Plant Efficiency Improvement project has yielded the following list of possible alternatives:

1. New power plant(s) would have been built.

2. Some of the power plants included in the project would have been retired.

3. The power plants included in the project would have been dispatched less frequently.

4. Various projects would have been undertaken by the SEBs with their share of the project funding, including, e.g., power plant availability improvement projects and T&D efficiency improvement projects.

5. Utility customers would have relied more heavily on self-generation.

Each of the above-identified alternatives must now be evaluated relative to the working hypothesis, and either rejected or accepted as the new hypothesis. We will begin with Alternative No 1. This alternative is based on the assumption that the efficiency improvements will result in an increase in power available to meet demand. Without this increase, it would be necessary to build new plants to meet demand.

Is such an alternative plausible? In a developed country, such as the United States, a heat rate improvement generally does not result in an increase in the total amount of generation available for final consumption. Here and in other industrialized countries, electricity supply and demand are in equilibrium; hence, efficiency improvements result in a reduction in the amount of fuel that must be consumed to meet load requirements, rather than an increase in the amount of electricity generated. However, in India the situation is much different, and an argument could be made that the efficiency gains realized through the project will result in an increase in total generation as well as a reduction in fuel consumption. At any given power plant, a portion of the heat rate reduction will be realized by improving the efficiency of auxiliary equipment, such as pumps and coal pulverizers. Given the supply-demand imbalance in India, the resulting reduction in-house power requirements should free up more generation for transfer to the grid. Furthermore, even improvements in boiler, turbine and condenser efficiencies may well result in increased generation as well as reduced fuel consumption.

Although turbine design constraints ultimately limit the amount of steam that can be passed through the turbine during any given time period, and hence, the total amount of generation, for a variety of reasons, power plants may not be able to operate at their maximum design steam load. For example, steam leaks from the boiler may limit the boiler's ability to provide steam to the turbine. Similarly, blockages in the condenser tubing may reduce allowable steam flow through the condenser, and hence, the turbine. Efficiency improvement efforts may correct these problems, thus enabling a return to operation at the turbine-generator maximum design output. In fact, the GEP project places great emphasis on condenser cleaning; along with boiler improvements, most of the efficiency gains from the project are expected to be realized through condenser cleaning efforts. Even if condenser cleaning does not improve steam flow, it should reduce the amount of power required for pumping, thereby freeing up more power for the grid.

In short, it is certainly plausible that a portion of the efficiency improvement resulting from the project will translate into a generation increase, at least up to the maximum design turbine-generator output. Once the plants have been returned to their design operating capacities, further efficiency improvements will reduce fuel consumption without increasing generation. To what extent a given heat rate improvement project may result in an increase in generation, as opposed to a decrease in fuel consumption, cannot be determined *a priori*; instead, tests must be performed (as shall be discussed in greater detail later in this chapter). However, certainly a baseline estimation procedure for the Indian project must accommodate the possibility of a significant generation increase, as well as a reduction in fuel consumption.

Figure 4.2. Establishing a Qualitative Baseline Under the Project-Specific Approach

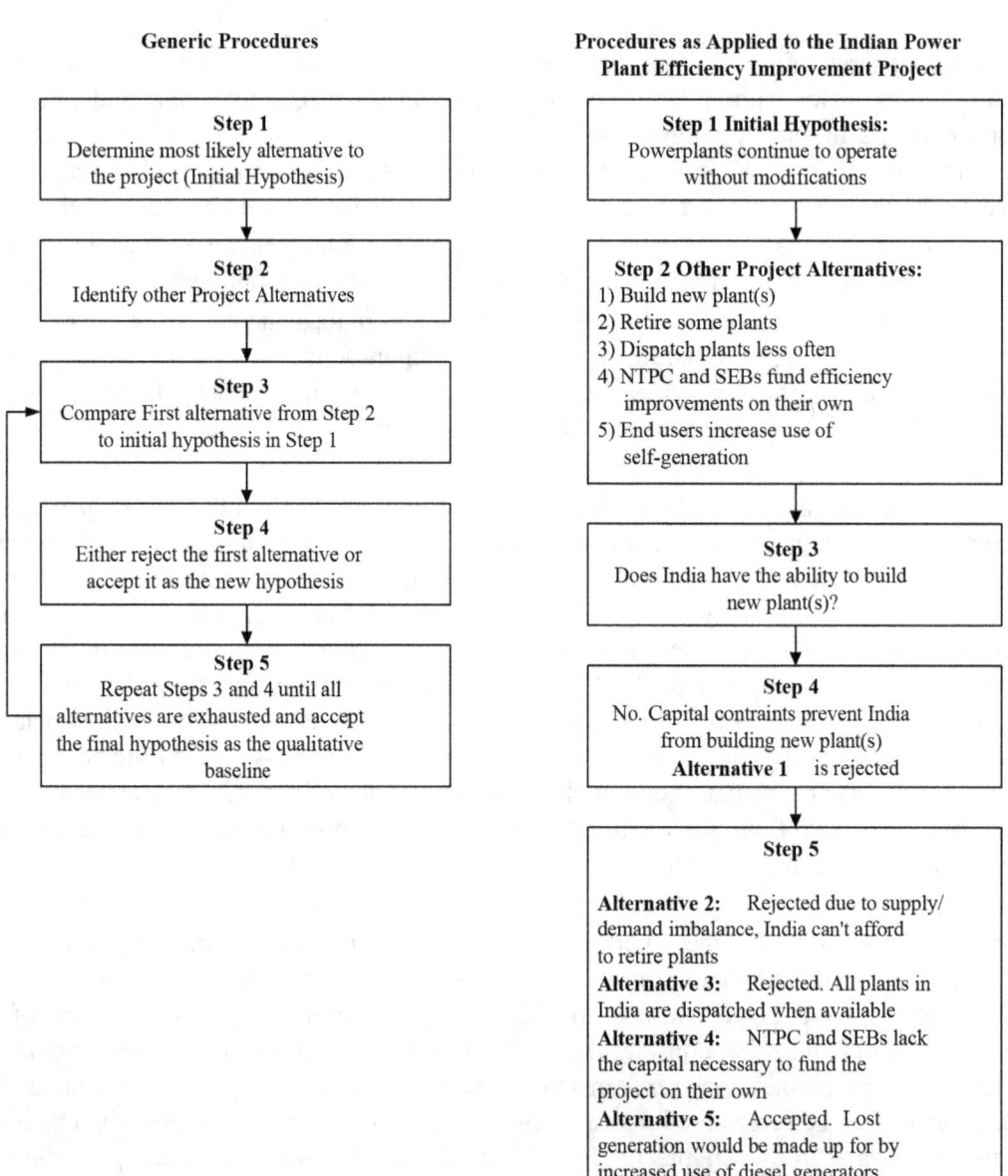

Generic Procedures

Step 1
Determine most likely alternative to the project (Initial Hypothesis)

Step 2
Identify other Project Alternatives

Step 3
Compare First alternative from Step 2 to initial hypothesis in Step 1

Step 4
Either reject the first alternative or accept it as the new hypothesis

Step 5
Repeat Steps 3 and 4 until all alternatives are exhausted and accept the final hypothesis as the qualitative baseline

Procedures as Applied to the Indian Power Plant Efficiency Improvement Project

Step 1 Initial Hypothesis:
Powerplants continue to operate without modifications

Step 2 Other Project Alternatives:
1) Build new plant(s)
2) Retire some plants
3) Dispatch plants less often
4) NTPC and SEBs fund efficiency improvements on their own
5) End users increase use of self-generation

Step 3
Does India have the ability to build new plant(s)?

Step 4
No. Capital contraints prevent India from building new plant(s)
Alternative 1 is rejected

Step 5

Alternative 2: Rejected due to supply/demand imbalance, India can't afford to retire plants
Alternative 3: Rejected. All plants in India are dispatched when available
Alternative 4: NTPC and SEBs lack the capital necessary to fund the project on their own
Alternative 5: Accepted. Lost generation would be made up for by increased use of diesel generators

80

In addition to the above considerations, it should be noted that improving the efficiency of existing power plants is identified as a top priority in the Government of India's current five year plan, perhaps in part because such improvements are a less-expensive way of increasing generation than building new power plants. Given this consideration, should we conclude that, without the potential for increased generation resulting from the project, India would have built more new power plants? There is no question that the *need* for new generating capacity might be greater in the absence of the project. However, the *means* for achieving that need would remain unchanged. Recall that although the SEBs are providing some of the resources for the project, their contribution is limited mainly to manpower and other resources in kind.[26] Therefore, the capital that would be freed up if the project were not undertaken would probably be insufficient to support the construction of new units. A lack of capital is in fact the key constraint that is preventing India from closing the supply-demand gap; this constraint will continue to exist regardless of whether or not the project is undertaken, and would effectively prevent the opening of new capacity to meet any generation shortfalls in the absence of the project.

Alternative No. 1, that new power plants would be opened in the absence of the project, is hence rejected. Alternative No. 2 is that all, or some, of the power plants included in the project would have been retired but for the project. This alternative is based on the assumption that the project will extend the life of some of the plants. However, as noted above, the project's goal is simply to return plants to their original design performance; life extension is not a project objective. Furthermore, none of the power plants included in the project are scheduled for retirement. Because of the supply-demand imbalance, India cannot afford to retire existing power plants, regardless of their age or efficiency. In fact, some Indian power plants are currently operating at efficiency levels below 15 percent owing to the great need for power. Hence, Alternative No. 2 can be readily rejected.

Similarly Alternative No. 3 can be rejected. This alternative assumes that the power plants included in the project would be dispatched less frequently in the project's absence. Certainly in the United States, an efficiency improvement at an existing power plant will reduce the plant's operating costs, which may in turn cause the plant to move up in the dispatch order. However, in India dispatching is much less of an issue. Given the large supply-demand imbalance, Indian power plants are generally dispatched whenever they are available, regardless of cost or efficiency. An improvement in the efficiency of some plants is therefore unlikely to have much effect on the dispatching of the plants.

Alternative No. 4 assumes that failure to undertake the project would free up capital within the SEBs, thereby enabling these organizations to pursue a variety of other generation, transmission and distribution projects. However, as we have seen in the argument rejecting Alternative No. 1, the monetary contribution of the Indian partners to the project is quite small. Hence, in the absence of the project any redirection of capital to other projects would be very limited if not insignificant.

---

[26]This statement, while true of our hypothetical project as we have defined it, is not true of the real project upon which it is based. The NTPC is actually providing $10 million towards the project, which represents the largest share of the project funding. We will return to consider the implications of this sizeable monetary contribution in Chapter 7.

Finally, in Alternative No. 5, we propose that, in the absence of the project, utility customers would have relied more heavily on self-generation. As a result of the poor reliability of power supply in India, many utility customers own backup diesel generators. In fact, the use of diesel generators is very common in the commercial sector, among both small and large establishments. Backup generation is also utilized in the industrial sector. However, in the residential sector, backup diesel generators are rarely found.

Diesel generators are used in part to backup the grid during grid failures. However, in addition to blackouts or brownouts caused by grid failures, India also experiences rolling blackouts used to manage the electricity supply-demand gap. When generation is insufficient to meet demand, rolling blackouts are in effect used to ration the available power supply. Diesel generators used during a rolling blackout (as opposed to a blackout caused by grid failure) are in effect serving to make up a portion of the supply-demand imbalance.

As we have already argued, the project may lead to an increase in the total generation available to the end users as well as a reduction in the fuel consumed for generation. This increased generation should in turn reduce either the length or areal extent of rolling blackouts. This being the case, it appears quite reasonable to conclude that, but for the project, diesel generators at the point of electricity consumption would be utilized more heavily. Diesel generation cannot entirely replace the increased generation that may result from the project. For one, the residential sector lacks backup generation capability, and hence, residential customers would simply be forced to reduce their electricity consumption in the project's absence. Furthermore, even in the commercial sector, the diesel generators would not replace all of the increased generation made available by the project. In many cases, the backup generators are manual start units; they do not switch on automatically when the power fails. Hence, these generators are presumably used only during normal business hours.

What may we conclude from the above considerations? First and foremost, the project may result in an increase in generation as well as a decrease in fuel consumption. Second, a portion of the increased generation sold outside the residential sector would displace small backup diesel generators owned by utility customers. The remaining portion in effect represents new generation that would not have been produced but for the project. Thus, it appears that our original hypothesis must be rejected in favor of Alternative No. 5. In the absence of the project, it is *not* true that nothing would have been done; instead, a portion of India's utility customers would have used their own diesel units to make up the lost generation. Furthermore, we may conclude that the project reduces emissions both at the affected power plants and at the backup generators. This follows from our conclusion that the project should result in increased generation up to the point of the plants' design capacities, and reduced fuel consumption beyond that point. Because the generation increase will be realized without an increase in fuel consumption, the project will result in real emission reductions at the end-use diesel generators, equal to the savings in diesel fuel times the diesel fuel emissions factor. Added to these reductions will be reductions at the power plants themselves, equal to the decrease in coal consumption times the emissions factor for coal. Having exhausted the list of possible alternatives to the project, we have now completed the generic procedure for establishing the qualitative baseline, and we have identified the baseline as the power plants and customer-owned diesel generators.

4.3.3.2 Quantifying the Baseline. Having identified the baseline in a qualitative manner, we must now quantify the baseline. Once again, we wish to propose a standard, step-by-step procedure for baseline quantification under the project-specific approach. This procedure is simple and straightforward, involving only three basic steps:

1. The algorithm(s) that will be used to quantify the baseline must be specified

2. The data inputs required to solve the algorithm(s) must be gathered or, when necessary, estimated

3. The algorithms must be solved.

Beginning with the first step, the proposed baseline estimation algorithms for the Indian Power Plant Efficiency Improvement project are as follows:

(1) $RDEC_j = [\Delta G_j - TD_j(\Delta G_j)][R_j(dr_j)(fr_j)]$

(2) $CDEC_j = [\Delta G_j - TD_j(\Delta G_j)][C_j(dc_j))(fc_j)]$

(3) $IDEC_j = [\Delta G_j - TD_j(\Delta G_j)][I_j(di_j))(fi_j)]$

(4) $ADEC_j = [\Delta G_j - TD_j(\Delta G_j)][A_j(da_j))(fa_j)]$

(5) $BE_j = (RHR_j)(RDEC_j)(DF) + (CHR_j)(CDEC_j)(DF) + (AHR_j)(ADEC_j)(DF)$
     $+ (IDF_j)(IDHR_j)(IDEC_j)(DF) + (IC_j)(ICHR_j)(IDEC_j)(C) + (FC_j + \Delta FC_j)(C)$

Where:

ΔG_j = Total change in net generation due to the project, in year j (MWh)

$RDEC_j$ = Estimated quantity of backup generation displaced in the residential sector, as a result of the project, in year j (MWh)

$CDEC_j$ = Estimated quantity of backup generation displaced in the commercial sector, as a result of the project, in year j (MWh)

$IDEC_j$ = Estimated quantity of backup generation displaced in the industrial sector, as a result of the project, in year j (MWh)

$ADEC_j$ = Estimated quantity of backup generation displaced in the agricultural sector, as a

83

result of the project, in year j (MWh)

TD_j = Fraction of total net generation lost due to T&D technical losses, in year j

R_j = Fraction of total generation delivered to the residential sector, in year j

C_j = Fraction of total generation delivered to the commercial sector, in year j

I_j = Fraction of total generation delivered to the industrial sector, in year j

A_j = Fraction of total generation delivered to the agricultural sector, in year j

dr_j = Fraction of residential electricity demand with backup generation capabilities, in year j

dc_j = Fraction of commercial electricity demand with backup generation capabilities, in year j

di_j = Fraction of industrial electricity demand with backup generation capabilities, in year j

da_j = Fraction of agricultural electricity demand with backup generation capabilities, in year j

fr_j = Fraction of rolling blackout time made up with backup generators, in the residential sector, in year j

fc_j = Fraction of rolling blackout time made up with backup generators, in the commercial sector, in year j

fi_j = Fraction of rolling blackout time made up with backup generators, in the industrial sector, in year j

fa_j = Fraction of rolling blackout time made up with backup generators, in the agricultural sector, in year j

BE_j = Project baseline emissions, in year j (tons CO_2)

RHR_j = Average heat rate of backup generators in the residential sector (mmBtus/MWh)

CHR_j = Average heat rate of backup generators in the commercial sector (mmBtus/MWh)

84

$IDHR_j$ = Average heat rate of backup diesel generators in the industrial sector (mmBtus/MWh)

AHR_j = Average heat rate of backup generators in the agricultural sector (mmBtus/MWh)

DF = emissions factor for diesel fuel (tons CO_2/mmBtu)

IDF_j = Fraction of industrial backup generator capacity that is diesel fired

IC_j = Fraction of industrial backup generator capacity that is coal fired

$ICHR_j$ = Average heat rate of backup coal-fired generators in the industrial sector (mmBtus/MWh)

C = Emissions factor for coal (tons CO_2/mmBtu)

FC_j = Total coal consumption at the project power plants, in year j (mmBtus)

$\triangle FC_j$ = Total change in coal consumption due to the project, in year j (mmBtus)

Equation 5 above yields the baseline emissions estimate, including both the estimated power plant emissions (absent the project) and the estimated emissions of the displaced backup generators. The project's *actual* emissions in any given year j would be equal to the project power plant emissions in that year. By subtracting these actual emissions from the baseline emissions computed in Equation 5, the creditable emission reductions would be derived.

4.3.3.2.1 Estimating the "Split" Between the Generation Increase and the Fuel Consumption Reduction. Some further explanation of the above equations is warranted. First, note that estimates of the generation increase ($\triangle G_j$) and fuel consumption decrease ($\triangle FC_j$) attributable to the project are required to solve the equations. Thus, it is necessary to determine not only the overall improvement in heat rate, but also to estimate what portion of this heat rate improvement goes towards increased generation, and what portion is used to reduce fuel consumption. Estimating this "split" between generation improvement and fuel consumption reduction will be a difficult exercise, requiring the development and implementation of power plant performance testing protocols. One possible approach would be to perform tests, prior to project initiation, to determine the actual (not the design) maximum turbine-generator output capability, under a variety of conditions (e.g., dry and monsoon seasons). Then, following project implementation, records of actual output on a short time interval basis (say, hourly) could be compared with the pre-project maximum capability; whenever the latter proves less then the former, the difference could be assumed to represent increased generation resulting from the project. By summing these hourly generation increases across all hours, the total generation increase resulting from the project could be estimated for each plant. The fuel consumption reduction could then be determined by solving the following equation for $\triangle FC_j$:

(6) $\Delta HR_j = (OFC_j/OG_j)-[(OFC_j-\Delta FC_j)/(OG_j+\Delta G_j)]$

Where:

    ΔHR_j = The overall change in heat rate due to the project, as determined through plant testing for year j (mmBtus/kWh)

    OFC = The plant's original (pre-project) fuel consumption used to estimate the pre-project heat rate (mmBtus)

    OG = The plant's original (pre-project) net generation used to estimate the pre-project heat rate (mmBtus)

    ΔG_j = The increase in the plant's net generation in year j, measured as described above (kWh)

    ΔFC_j = The reduction in the plant's fuel consumption in year j (mmBtus)

The above equation can be rewritten and solved for ΔFC_j once the increase in net generation has been determined:

(7) $\Delta FC_j = \{(OG+\Delta G_j)\ [\Delta Hr_j-(OFC/OG)]\}-OFC$

Note that Equations 1 through 5 do not assume a significant increase in generation resulting from the project; even if it is determined that there is little or no generation increase the equations will still apply. Rather, the equations are designed to accommodate *any* generation change, ranging from insignificant to substantial. Thus the equations require no *a priori* assumptions in this regard; they are applicable regardless of the outcome of the tests described above.

4.3.3.2.2 The Treatment of Time in the Equations. Secondly, note that most of the variables in the equations have a time dimension; hence, the specific values of these variables will change from year to year throughout the life of the project. In this way, temporal change in the most likely project alternative is incorporated directly into the baseline equations. Of particular importance, notice that the various sectoral electricity consumption fractions (e.g., C_j, dc_j, and fc_j for the commercial sector) all have a time subscript. Thus, changes in the market penetration of backup generators (dc_j for the commercial sector), as well as changes in their utilization (fc_j for the commercial sector) and changes in the overall sectoral distribution of electricity consumption (C_j for the commercial sector) will be captured by the algorithms. Improvements in backup generator efficiencies will be captured by the various heat rate variables in Equation 5, all of which have a time dimension. Similarly, variations in heat rate, generation and fuel consumption levels at the power plants included in the project will be captured by the time-subscripted variables ΔG_j, FC_j, and ΔFC_j.

Equations 1 through 5 will also capture variations in the magnitude of India's supply-demand imbalance. At present, a significant electricity supply shortfall exists, and it is likely that this shortfall will continue to exist for some time in the future. However, the project is expected to have a very long life, as it is extended to more and more power plants operated by the SEBs. It is therefore certainly plausible, if not probable, that the supply-demand gap will be closed within the project's lifetime. If the supply-demand gap gradually diminishes, this will manifest itself in the equations as a gradual reduction in the value of ΔG_j (the change in net generation) and a commensurate increase in the value of ΔFC_j. Eventually, the value of ΔG_j may reach zero; at this point the supply-demand gap will have closed and further reductions in emissions will occur solely as a result of reductions in the level of fuel consumption at the power plants.

4.3.3.2.3 Simplifications Incorporated in the Algorithms. Third, it is important to note that the algorithms presented above represent simplifications of the ideal or "true" algorithms. Three main types of simplifications were made in developing the algorithms. Perhaps the most important simplifications have to do with the estimation of the amount of backup generation displaced by the project in Equations 1 through 4. These four equations utilize sectoral averages. But in the ideal, these equations would be replaced with a separate equation for every backup generator affected by the project. Similarly, the sectoral terms in Equation 5 would, in the ideal, be replaced with separate terms for each generator. Instead of representing each individual generator in the equations, it is necessary to utilize a sectoral representation, with estimates of sectoral averages for heat rates and other variables, owing to data limitations. The backup generators affected by the project number in the tens, if not hundreds, of thousands; furthermore, these generators are scattered throughout India, and in many cases are located at very small commercial establishments. Public data on the vast majority of India's backup generating units are simply nonexistent, and clearly the cost of doing a comprehensive survey of these generators would be prohibitive.[27]

The second set of simplifications has to do with the treatment of time in the equations. Note that the subscript j represents a specific *year* for which the baseline is required. In the ideal, a much finer time increment – e.g., an *hour* – would be utilized. This, combined with the use of separate equations for each backup generator, would enable a more accurate representation of backup generator utilization. Specifically, it would be possible to identify the specific areas of India affected by outages at any particular point in time, and to estimate more accurately, based on the time of day, whether or not the backup generators in those areas would be utilized to make up the lost power. The hourly equations would be solved for each hour comprising a given year, and the resulting hourly baseline emission estimates would be summed across all hours to yield the annual total. However, given the need to use sectoral averages rather than individual backup unit data, the accuracy gains resulting from such an approach would be limited and were judged to be insufficient to warrant the enormous added

---

[27]In the context of a real as opposed to a hypothetical market-based project, the cost of doing a survey of a small sample of the generators might not be prohibitive. Furthermore, the costs of such a survey to project developers could be effectively reduced if the costs could be shared among a number of market-based projects.

complexity and expense that would be involved.[28]

In addition to aggregating data across sectors and across time, the algorithms incorporate a third set of simplifications: they ignore the project's secondary effects. Given the current supply-demand imbalance in India, and the consequent need to dispatch all power plants whenever they are available, it is unlikely that this project will have a significant impact on dispatch orders. However, because the project is expected to affect power plant fuel consumption, it may also affect coal mine methane emissions as well as the emissions associated with transporting coal from the mines to the power plants. In addition, reductions in backup generator use should reduce India's overall demand for diesel fuel, with consequent reductions in the emissions associated with fuel production, processing, and transportation.

There is another *possible* secondary effect, of far greater consequence not only for this particular project but for the concept of market-based activities on a whole. Many developing countries – including China, as well as India – experience significant fuel supply bottlenecks. In India, coal supply is constrained by a number of factors, including regional mismatches in supply and demand, volatility in the quality (i.e., Btu content) of coal shipments, rail transportation bottlenecks, and difficulties and delays in opening new mining capacity. Capital projects to increase capacity have been discouraged by government-controlled coal pricing policies; as a result, of the 72 million metric tons of new capacity to have been sanctioned under the Eighth Five-Year plan, only 43 million metric tons of capacity actually was sanctioned. Railroad improvement projects have also been delayed owing to capital constraints; as a result, mine-mouth coal stockpiles have grown even though Indian power plants must often operate with inadequate on-site fuel stocks.

As a result of these problems, a 20-25 million metric ton coal supply shortfall for the power sector materialized in the 1996-97 period, according to the government's Ninth Five-Year Plan. The same plan projects a 20.3 million metric ton shortfall in 2001-02. To put these numbers in perspective, the GEP project is expected to save 9 million metric tons of coal annually, once the project has been extended to all of India's thermal power plants. Furthermore, the total coal handling capacity at India's major ports is only 8.5 million metric tons per year, thus limiting the country's ability to make up the expected shortfall by increasing imports.

In light of the supply constraints, it is certainly plausible, if not likely, that any coal saved as a result of the project would simply be used either by the power sector itself to increase generation, or by other sectors of the economy. Because coal supply is not matched to demand, India's power plant coal consumption is determined by its capacity to mine and move coal to the plants. If demand is reduced even by as much as 10 million metric tons supply and consumption may remain constant simply because, despite the reduction, demand may *still* exceed supply by about 10 million metric tons. The implication, for our project, is that the reductions in fuel consumption, and emissions, at the power

---

[28]In the context of a real as opposed to fictitious market-based project, the additional expense might be justified.

88

plants might not materialize.[29] However, the implications are in fact far broader than this, because the above arguments would appear to apply to *any* project designed to reduce coal consumption in India, and, for that matter, to *any* project designed to reduce fuel consumption in a country facing significant fuel supply-demand imbalances. Such imbalances are likely to be quite common in the developing world, as a result of such factors as capital constraints, inadequacies in transportation infrastructure, and rapid growth in GDP and electricity demand.

However, the situation is more complex than this overview suggests, in India itself and elsewhere. First, the fuel supply-demand gaps are not constant over time, but vary significantly depending on numerous factors such as coal mine productivity, coal quality fluctuations, competing demand for rail services, weather, etc. Second, a reduction in demand at one power plant will not necessarily "free up" coal for shipment to another consumer. Rail capacity constraints may limit the country's ability to reroute coal shipments from the project power plants to alternative consumers. In short, a detailed analysis would be required to determine the extent to which coal consumption reductions might be offset by secondary effects. Such an analysis is beyond the scope of this report. However, given the potential importance of secondary effects arising from fuel supply constraints in developing countries, we will return to this issue in the concluding chapter (Chapter 7).

4.3.3.2.4 Miscellany. A few other miscellaneous points should be made concerning the algorithms. First, note that in Equation 5, separate terms are included for diesel- and coal-fired backup generators in the industrial sector. Our *a priori* assumption is that both diesel and coal-fired generation is utilized in the industrial sector, but that the other sectors rely strictly on diesel generators. Obviously if actual data are obtained indicating the use of coal-fired generation in other sectors, or, for that matter, natural gas or other types of generation, Equation 5 can and should be modified to represent these other fuel types. Second, note that the variable representing transmission and distribution (T&D) losses in Equations 1 through 4 includes only technical losses from system inefficiencies. Non-technical losses (i.e., non-payment, theft, or subsidized payment) are excluded, because such losses represent a misallocation, rather than an actual loss, of electrical energy. For purposes of estimating the emission baseline, we are not concerned with the issue of whether or not individual end-users are paying for the electricity they consume. Our exclusive concern is whether or not end-users are using backup generators in the event of power outages or shortfalls and, presumably, end-users with illegal connections to the grid may be using backup generators just as paying customers use them.

The actual emissions of the power plants in any given year j would be measured and subtracted from the baseline emissions computed in Equation 5, to yield the estimated emission reductions. Credits equivalent to these estimated emission reductions would be awarded to the project developers on an annual basis.

---

[29]Reductions in emissions on a per kilowatt-hour basis would still occur; however, credits can only be awarded on the basis of an absolute reduction in emissions. Absolute reductions in the emissions of diesel generators would still occur regardless of the fuel supply shortfalls.

4.4 Summary

In this chapter, we have applied the project-specific approach to a hypothetical project representing an extension of an actual, ongoing project: India's GEP project. This project essentially involves a systemic approach to improving the heat rate of India's coal-fired power plants. A series of diagnostic tests are performed on the boilers, turbines, condensers, at each power plant. The tests enable identification of specific plant components that are operating at less than design or optimal efficiency. The tests also provide some indication of the corrective actions required to improve component performance. These actions, which for the most part involve procedural changes or limited capital improvements, yield efficiency gains which in turn reduce greenhouse gas emissions.

Because the project is designed to yield significant heat rate improvements with only modest cost outlays, it would be economically viable even without the financial aid provided by the project's U.S. sponsors. Hence, the additionality of the project could not be demonstrated utilizing economic feasibility analysis techniques. Instead, the non-economic barriers approach was used to demonstrate additionality. Specifically, it was argued that the project could not have been undertaken without the support of the U.S. sponsors, because of the existence of two project barriers: a financial barrier, which constrains the capital available to fund such projects in India; and a knowledge barrier, which would have prevented India's electric utilities from undertaking the project without U.S. technical expertise. Based on these arguments, the project's additionality was deemed to have been demonstrated for the first five years of the project life. For this project, it was recommended that additionality be re-evaluated on a five year basis, based on up-to-date information concerning the financial conditions of India's electric utilities.

Finally, the emissions baseline for the project was identified and developed. As a result of a significant electricity supply shortfall in India, it was recognized that heat rate improvement projects, such as this one, may lead to an increase in net generation as well as a reduction in fuel consumption. Hence, in addition to reducing emissions at the affected power plants, the project may also reduce the utilization of backup diesel generators, used by commercial and industrial establishments throughout India during the rolling blackouts necessitated by the supply shortfall. A set of algorithms (Equations 1 through 7 above) were developed to estimate baseline emissions at both the power plants and the end-use backup generators.

In Chapter 7, we will return to the Indian Power Plant Efficiency Improvement Project, to present our critique of the analysis developed in this chapter. First, however, we turn to an analysis of a hypothetical IGCC project in China, using the modified technology matrix approach.

5. EMISSIONS BASELINE DEVELOPMENT FOR AN INTEGRATED GASIFICATION COMBINED CYCLE (IGCC) POWER PROJECT IN CHINA

5.1 Project Description

5.1.1 Background

The market-based project using Integrated Gasification Combined Cycle (IGCC) technology in the People's Republic of China (PRC) has been selected in response to Chinese government plans to build an IGCC demonstration project and to develop domestic capability to produce IGCC technology. As no IGCC plant has yet been commissioned in China, the information used in this chapter is based on data and experience derived from existing IGCC plants in the US and Europe.

Coal is the dominant source of energy supply in the PRC, as a result mainly of the abundance of domestic resources of coal relative to oil and natural gas. In 1999, coal provided 73 percent of total energy consumed in China while oil provided 20 percent, natural gas provided 2 percent, hydro provided 5 percent, and nuclear provided 0.4 percent of energy consumed. The heavy reliance on coal has caused serious environmental impacts at every stage of the coal chain. In 1997, for example, carbon dioxide emissions in the PRC reached 821.77 million metric tons of carbon, second only to the carbon dioxide emissions of the U.S. In 1996, 83 percent of carbon dioxide emissions came from coal combustion.[30] Energy sector pollution has been magnified by the high intensity with which energy has been utilized in China. The average thermal efficiency of China's power plants is 25 to 29 percent compared to rates of 35 to 38 percent in industrialized countries.[31] Over half of the existing power plant capacity is in units below 200 MW.

During the last 15 years, annual growth in electricity generation averaged 8 percent, yet in most regions, electricity supply did not keep pace with growth in demand. Following the Asian financial crisis, the Chinese electricity sector has experienced a slow-down in demand, leading to an over-supply of coal and electricity in some regions.[32] The Chinese government has therefore imposed a temporary moratorium on the approval of new power plants and is implementing a drive to close down small and highly polluting thermal power plants. For example, the Chinese government has ordered the shut down of a number of old, small (<100 MW), inefficient coal-fired power plants, totaling about 14 GW of generating capacity. However, in the longer term, electricity demand is expected to revert back to its previous growth rate of about 8 percent per annum. More than 90 percent of the investment needed to satisfy this demand growth is likely to go towards coal-based power generation. Therefore, to reduce carbon emissions, improve overall energy efficiency, and reduce fuel use, the Chinese

---

[30] "Coal in the Energy Supply of China." Report of the CIAB Asia Committee. Coal Industry Advisory Board (CIAB), International Energy Agency (IEA). Paris, France 1999.

[31] Blackman, Allen and Xun Wu. Foreign Direct Investment in China's Power Sector: Trends, Benefits and Barriers. Resources for the Future. Washington D.C. September 1998.

[32] U.S. Energy Information Administration (EIA), June 1999. http://www.eia.doe.gov/emeu/cabs/china.html

government has emphasized the need for adopting, producing and marketing cleaner coal technologies, including IGCC technology.

5.1.2 The Project

The construction of a commercial scale demonstration IGCC plant by 2000 has been listed as a priority under the PRC's Agenda 21 program. Beijing No. 3 Thermal Power Plant and the Yantai Thermal Power plant in Shangdong have been considered as potential sites. In October 1997, the Ministry of Electric Power (now the State Power Corporation) selected Yantai in the Northeastern coastal province of Shangdong as the site for China's first IGCC demonstration plant and applied to the State Planning Committee for the approval to establish the plant.[33] However, the status of the Yantai project in terms of government approval and project financing is still unclear.

Based on this interest in advancing IGCC technology in China, we developed a sample market-based project that would rely on the construction of an IGCC power plant in China. This project could either be implemented in the city of Yantai, which has already expressed an interest in establishing an IGCC power plant, or it could be implemented in some of the inland provinces facing serious electricity shortages, including Gansu, Henan, Wuinghai and Sichuan. The hypothetical project would consist of two units, each adding between 300 to 400 MW of new generating capacity to the grid, and the targeted unit efficiency would be 43 percent or higher. The project would rely on imported technology obtained through a combination of direct purchase and technology transfer. Financing for the project could be obtained from international lending institutions, such as the World Bank and the Asian Development Bank, and from American and European investors who have already expressed interest in providing financial support for this type of projects in exchange for emissions credits. The amount of emission credits awarded to each individual investor would be determined by the share of financial assistance provided by each one of them.

IGCC technology combines coal gasification with a combined-cycle power plant. The first step in the IGCC process is coal gasification whereby solid coal is reacted with steam and oxygen to produce a combustible gas composed primarily of carbon monoxide and hydrogen, also known as syngas. To support the coal gasification process, several additional plant sections are needed. An air separation unit (ASU) produces oxygen for the gasification process, while the impurities in the raw gas stream from the gasifier are removed in either a hot gas or a cold gas clean-up plant. After the gas has been cleaned of sulfur compounds and particulates, it is burned in a gas turbine to generate electricity. Meanwhile, exhaust gas from the gas turbine passes through a heat recovery boiler to produce steam that drives a separate steam turbine generator providing an additional source of electricity.

---

[33] The Nautilus Institute for Security and Sustainable Development. "IGCC in China". A background paper for the ESENA Workshop on Innovative Financing for Clean Coal in China: A GEF Technology Risk Guarantee? Berkeley, California. February 27-28, 1999, and Jiang Zhesheng, China State Electric Power Corporation. "IGCC Demonstrated Power Plant in China". Presented at the ESENA Workshop on Innovative Financing for Clean Coal in China: A GEF Technology Risk Guarantee? Berkeley, California. February 27-28, 1999.

Compared to the conventional pulverized coal (PC) power plants that are currently used widely in China, the proposed IGCC power plant would offer several new features, including the gas cleanup process which purifies the syngas of sulfur and particulate pollutants before combustion in the turbine, and the utilization of the residual heat from the hot exhaust gas to produce additional electricity. Because the pollutants are removed before the fuel is burned in the gas turbine, smaller volumes of gas need to be treated as compared to a post-combustion flue gas desulfurization (FGD) device. However, the gas stream must be cleaned extremely well to achieve low emissions and to protect downstream components, such as the gas turbine, from corrosion and erosion.

The IGCC market-based project would also provide several environmental benefits. First and foremost, IGCC technology improves thermal efficiency considerably compared to conventional coal-fired power plants. Similar IGCC systems in the U.S. and Europe have already reached conversion efficiencies of 43 to 47 percent, compared to 36 to 38 percent for a new conventional subcritical PC plant. Fuel use therefore will be reduced and should help alleviate some of the bottlenecks associated with transporting coal on the congested rail system from the major mine fields in the Chinese far North. Second, the IGCC plant has the potential to reduce SO_2 by 95 to 99 percent, NO_x by 90 percent, particulates by 98 percent and carbon dioxide (CO_2) by 34 percent compared to a traditional PC plant. Among other goals, the project is targeted to reach a desulfurization efficiency of no less than 98 percent per generating unit. This is important because most of the coal found in China has a high sulfur content. As a third benefit, the IGCC plant will be capable of using any high hydrocarbon fuel, including low- and high-sulfur coal, anthracite, and biomass, adding an unprecedented measure of flexibility to switch fuels as costs change. The IGCC plant will also produce several usable by-products with its solid waste, including inert vitreous slag, which has a number of applications in the construction industry, and recoverable sulfur, which can be marketed either as pure sulfur or sulfuric acid. Finally, the IGCC project will result in significant savings in water usage as the water required to operate an IGCC plant is only 50 to 70 percent of that required to run a PC plant with a FGD system.

The IGCC project will be constructed step by step by building the gas turbine first, then converting the plant to a gas-fired combined cycle plant, before finally completing the IGCC plant by adding the gasifier. This approach offers the advantages of short construction time, low initial investment, high efficiency, and low pollution that is crucial in an area with rapid economic growth and growing electricity needs. According to the Chinese Government, the long-term objective of building an IGCC power plant is to replicate this technology throughout China. By testing imported technology on Chinese soil, the Government intends to encourage other regions to build their own commercial scale IGCC power plants and eventually develop the capability domestically to assemble and construct such plants. The expectation is that once China is able to produce its own IGCC technology, construction costs will fall, the same way the Chinese managed to bring down costs by constructing their own pulverized coal plants.[34]

---

[34] Nautilus Institute "Financing Clean Coal Technologies in China". Background paper for the ESENA Workshop. February 27-28, 1999.

5.2 Emission Baseline Development

5.2.1 Evaluation of Baseline Options

To estimate the emissions reductions for the IGCC project, we first have to select a baseline development approach. As noted in Chapter 3, two generic approaches can be used for baseline development; the modified technology matrix approach and the project-specific approach. As noted in Table 3.1 of Chapter 3, Type 1 projects involving the installation of new capacity and utilizing advanced qualifying technologies should apply the modified technology matrix approach. The Chinese IGCC project clearly falls into this category. Although IGCC technology has been on the market for more than a decade, only five commercial scale coal-fired IGCC power plants have been constructed worldwide and none of these are in a developing country. As an emerging technology not yet available in China, IGCC technology would obviously qualify for inclusion in China's modified technology matrix.

In addition, the Type 1 approach should be utilized because the Chinese IGCC sample project involves the construction of new generating capacity. As noted in Chapter 3, most projects involving the opening of new capacity will have to use some sort of benchmark to estimate the emissions baseline. For this purpose, the modified technology matrix is very useful because it introduces a level of standardization and objectivity into the estimation process and reduces baseline estimation costs for the project developers.

In this chapter, we will focus on *qualifying* and *developing* the stipulated benchmark for inclusion in the modified technology matrix, rather than *applying* the baseline method to the particular project for the purpose of estimating the exact emission reductions achieved. Normally, once a market- or project-based activity is implemented, a project developer wishing to apply the modified technology matrix baseline approach would simply verify that the particular technology was included in the matrix and represented new generating capacity. The project developer would then use the specific benchmark assigned to the technology in order to calculate the emission reductions of the project. At this point however, the list of pre-qualifying technologies has yet to be developed. The process of qualifying a specific technology for inclusion in the matrix is rather complex and requires using both the additionality test and the development of an appropriate emissions benchmark for the technology. This effort raises several interesting questions and entails considerably more work than the actual application of the modified technology matrix as a baseline approach.

Hence, the rest of this chapter will focus on qualifying IGCC technology for the modified technology matrix by demonstrating additionality and developing the stipulated benchmark. This will be done by analyzing IGCC technology in general without reference to a specific project. As the stipulated benchmark should be applicable for the evaluation of other IGCC projects implemented in China, it cannot be based on a single project. Rather, to ensure objectivity, the baseline should be derived on the basis of the qualifications of the technology itself.

Finally, it should be noted that the following analysis only refers to *coal-fired* IGCC power plants. IGCC power systems allow for a variety of fuel-uses (biomass, coal, oil, and waste) and can be applied for different purposes, such as cogeneration, refining, and production of chemicals. Moreover, IGCC technology can easily be used for repowering existing power plants. As these uses and applications may result in vastly different emission rates and cost estimates than those associated with a coal-fired IGCC unit, projects applying these technologies will not be able to apply the emissions baseline developed in this chapter. Rather, these variations of the IGCC technology would have to be evaluated separately for their applicability with respect to the technology matrix and would eventually be listed as different entries on the list of qualifying technologies.

5.2.2 Additionality

The next step in the process of developing a Chinese IGCC market-based project involves demonstrating the additionality of the technology. As established in the previous section, the modified technology matrix approach will be used for developing the emission baseline for the IGCC project. Hence, there is no need to establish additionality on a project-by-project basis. Instead, the application of stringent criteria to qualify the technology for inclusion in the modified technology matrix will ensure the additionality of the project. In close cooperation with individual host countries, those directly involved with administration of international market-based activities will be responsible for undertaking this technology evaluation and qualification, before actual projects will be developed. Once this has been accomplished, all project developers will need to do to establish additionality, is to provide evidence that their project utilizes a qualifying technology (in this case IGCC) from the technology matrix. However, as the list of pre-qualifying technologies has not yet been developed, this section will proceed to evaluate IGCC technology for the purpose of adding it to China's technology matrix.

How stringent should the criteria be that are used to determine inclusion in, or exclusion from, the technology matrix? The answer to this question must be based on a consideration of the tradeoff between costs, on the one hand, and accuracy on the other. For the purpose of the case studies, our goal is to minimize errors as much as possible to those that may be inherent in the baseline development approach, and to exclude errors that may result simply form lack of diligence. Hence for present purposes the criteria used to demonstrate the additionality of a technology under the modified technology matrix approach should be very stringent, to ensure that only projects proven to be non-commercial in the absence of the project- or market-based activity receive credit. As the additionality test evaluates a technology in general, and not the individual projects utilizing this particular technology, a less stringent set of evaluation criteria could potentially allow a large number of non-additional projects to receive credit, thus encouraging exploitation and bias.

The additionality of a given technology may be demonstrated by evaluating the *economic feasibility* and the *market penetration* of the technology. If the technology is found to be unable to compete economically with existing technologies on the market and has failed to reach even a minimal level of market penetration, the technology will qualify as additional. This evaluation should be based on the specific circumstances of the host country. Ideally, the technology should qualify only if it meets both

95

the economic feasibility and the market penetration test. As will be illustrated in the following pages, IGCC technology passed both tests and therefore can be included in the modified technology matrix.

5.2.2.1 Economic Feasibility. To demonstrate economic additionality, IGCC technology must be evaluated for its economic feasibility; that is, the cost of building an IGCC power plant should be compared to the cost of building another power plant employing alternate technologies. If the IGCC power plant is found to be economic without the favorable financing provided by the market- or project-based activity, it will not pass as additional under the economic feasibility test.

Over the past two decades, numerous research programs and demonstration projects have been implemented to develop and commercialize IGCC technology. Furthermore, there are five commercial scale coal-fired IGCC plants in operation in the world. Three of these projects, the Wabash River Coal Gasification Power Plant, Tampa Electric Company's IGCC Project, and Piñon Pine IGCC Project, are in the U.S. and have been implemented with financial support from the U.S. Department of Energy's Clean Coal Technology Program. The two most recent plants are in Europe; the Puertollano coal- and petcoke-fired IGCC plant in Spain and the coal-fired IGCC plant in Buggenum, The Netherlands. The American plants use General Electric (GE) gas turbines and have reached an average efficiency of around 40 percent. Both European plants use Siemens gas turbines. In 1998, the Buggenum plant reached an efficiency of 43 percent while Puertollano demonstrated a 47 percent efficiency rate.

In spite of these developments, IGCC technology is still not viewed as a proven or mature technology, particularly not in developing countries. The construction costs and construction time of an IGCC plant cannot be predicted with certainty, and the plant's operational performance, such as plant availability, cannot be guaranteed either. Most importantly, IGCC capital costs are higher than the costs of building a conventional coal-fired power plant. As illustrated in Table 5.1, the capital costs of building an IGCC power plant currently range between 1300 to 1350 $/kW as compared to 1000-1200 for a conventional pulverized coal plant with flue gas desulfurization (FGD) controls.[35] In fact, IGCC technology is currently the most expensive option of all the advanced coal technologies available today, including atmospheric fluidized bed combustion (AFBC) technology and pressurized fluidized bed combustion (PFBC) technology.

IGCC technology becomes even less economic when compared to a natural gas-fired power plant. Presently, a CT/CC power plant costs 400-500 $/kW to build, representing a third of what it would cost to build a coal-fired IGCC power plant. However, it should be noted that natural gas prices are considerably higher in China owing to the scarcity of natural gas outside the coastal areas. The disparity in terms of price between a natural gas and a coal-fired power plant in China therefore would not be as pronounced as the numbers in Table 5.1 indicate.

---

[35] Mollot, Darren J. USDOE. "Clean Coal Technologies & Beyond: The Status and Prospects of Coal Power" Presented at the Johns Hopkins University, School of Advanced International Studies (SAIS). April 28, 1999.

Table 5.1. Comparative Economic Factors for U.S.-Based Power Generation Systems.

System	Average Efficiency - HHV (%)	Capital Costs ($/kW)
Subcritical PC w/o controls	38	800-1000
Subcritical PC w/ controls	36	1000-1200
Supercritical PC	43	950-1600
AFBC	36	950-1150
PFBC	42	1250-1350
IGCC	42	1300-1350
CT/CC	52	400-500

Source: Darren J. Mollot, U.S. Department of Energy (1999); and Coal Industry Advisory Board, International Energy Agency, Paris, France (1998).

When looking at the costs of constructing an IGCC plant in China, the high costs of the technology become even more pronounced. The Nautilus Institute estimates that a 500 MW IGCC project in China would generate power at a cost of ¢5.8/kWh while a conventional coal-fired power station would supply power at a cost of ¢4.4/kWh.[36] This price difference is based on the cost figures provided in Table 5.2. According to the Nautilus Institute, these figures imply that the IGCC plant would cost some $70-80 million more to implement than a conventional pulverized coal plant. However, it should be emphasized that the environmental performance of IGCC technology is far better than that of conventional PC-fired power plants in China. Moreover, as noted in Table 5.1, the cost difference would be less pronounced if IGCC were compared to imported supercritical PC technology, which is the current state-of-the art technology.

Table 5.2. Levelized Costs for IGCC Versus PC in China.

	PC w/FGD (US)	PC w/FGD (China	IGCC (US)	IGCC (China)
Capital Costs ($/kW)	1,150	880	1,450	1,500
Construction Time (yrs.)	5.0	5.0	5.0	5.0
Thermal Efficiency (%)	35	32	44	41
Plant Availability (%)	74	70	70	70
Delivered cost of power generation (¢/kWh)	4.7	4.4	5.7	5.8

Source: Nautilus Institute (1999)

However, capital cost is an issue that will greatly influence the speed at which China adopts IGCC technology. Even though IGCC is the cleanest most efficient coal-fired technology available today and its capital costs are expected to decline, it will not be widely utilized if it is not affordable for China. The country is only beginning to build the capability to manufacture, build, or operate IGCC

---

[36]The Nautilus Institute. "Proposal for a Global Environment Facility (GEF) Technology Risk Guarantee Mechanism." A background paper for the ESENA Workshop on Innovative Financing for Clean Coal in China: A GEF Technology Risk Guarantee? Berkeley, California. February 27-28, 1999.

facilities.[37] As a result, the first demonstration projects are likely to be high in cost, even though labor costs are lower in China. The reluctance to adopt IGCC technology is exacerbated by the fact that China lacks the current production capacity and financial wherewithal required to both keep up with current demand growth and replace some of its oldest and most polluting power plants. For example, Chinese power authorities estimated that China would only be able to domestically finance 80 percent of the investment needed to meet China's year 2000 capacity target.[38] As a coal-fired steam turbine power plant with all in-country content can be built in China at a cost lower than any other country in the world, it is not likely that the financially constrained Chinese power authorities will readily turn towards the much more expensive IGCC technology.[39]

Another issue, which influences the cost-advantage and therefore the rate of adoption of IGCC technology in China, is environmental regulation and enforcement. For years, the central government did not pay particular attention to controlling emissions but instead focused on meeting the rapidly growing electricity demand as quickly and cheaply as possible. However, recently the government imposed more stringent environmental regulation announcing the intention to keep particulate emissions below 3.8 million tons per year and SO_2 emissions below 15 million tons per year.[40] As a result, the use of emission controls such as FGD are becoming more widespread.

However, the regulations, and the ensuing penalties for non-compliance, have not in the past been stringent enough to create an economic incentive for adopting more advanced clean coal technologies, such as IGCC. In fact, many cash-strapped local governments continue to select the type of plants that have low, up-front costs, even though they may be inefficient and involve a higher operating cost and life-cycle costs.[41] In some cases, investors have even built plants that they knew would not meet specified emissions standards because they preferred to pay the emission penalties later rather than increasing up-front costs. The push-factor towards implementing IGCC technology will not become significant until the Chinese government raises pollution standards and penalties for non-compliance and/or introduces reforms that improve the cost and availability of public and private financing.

Based on an economic comparison of IGCC technology with other coal-fired technologies on the market, it becomes evident that IGCC technology is not commercial in China at the moment. Unless favorable financing is provided through innovative mechanisms such as market- or project-based activities IGCC technology will continue to be viewed as too expensive and is not likely to be

---

[37]Currently, China has experience in operating gasifiers for town gas and chemicals production. As a move towards developing domestic capability to use IGCC technology, Texaco recently licensed the Jinzhou Heavy-Duty Machine Plant (JHDMP) in Dalian, Liaoning Province to be the first Asian qualified provider of Texaco pressure vessels and spare parts for IGCC.

[38]Blackman, Allen and Xun Wu. "Foreign Direct Investment in China's Power Sector: Trends, Benefits and Barriers." Resources for the Future. Discussion Paper 98-50. September, 1998.

[39] Studies indicate that the cost of constructing a PC plant in China can be 32 percent less expensive than in the U.S. mainly because of differences in costs of labor and raw materials. Nautilus Institute "Financing Clean Coal Technologies in China". Background paper for the ESENA Workshop. February 27-28, 1999.

[40] Ibid.

[41] Ibid.

introduced in China on a wide scale – at least not in the near future. The economic feasibility test thus indicates that IGCC technology is additional and qualifies for inclusion in the modified technology matrix.

5.2.2.2 Market Penetration. A second additionality test, based on market penetration, can be applied to account for other non-economic barriers to implementation that are more difficult to measure and verify. Such barriers may include risks associated with installing and operating locally unknown technologies, institutional or trade barriers, or internal organizational structures that discourage investment in energy sector improvements. The notion is that these barriers may prevent otherwise economically healthy technologies from being implemented without receiving government funding or other subsidies, that is, they would be unable to gain even a minimal market penetration rate in a given host country. As a result, these technologies would qualify as additional and be included in the technology matrix.

In the case of IGCC technology, it is fairly straightforward to establish the market penetration rate and demonstrate additionality, because no coal-fired IGCC power plants have been built in China to date. To indicate the level of barriers facing the implementation of IGCC technology in China, it should be noted that the level of commercial risk facing the construction of a 500 MW IGCC plant is estimated to be 23 percent higher than that of a conventional power plant of equal size.[42]

To allow optimum penetration of a technology in China, a significant portion of the plant equipment would have to be manufactured domestically. As it turns out, China does have most of the capacity to manufacture IGCC units domestically. Over the years, China has developed the capability to produce gas steam combined cycle units up to a size of 600 MW. The Chinese also have some domestic experience with manufacturing coal gasification units. China has numerous domestic gasifiers for town gas and chemicals production. In fact, JHDMP has equipped more than 10 companies with gasifiers and has a capacity to produce more than a dozen units annually. Thus, the biggest problem facing the Chinese will be combining the various units to successfully operate an IGCC power plant.

Even on a worldwide basis, coal-fired IGCC technology is only slowly gaining ground. Of the five commercial scale coal-fired IGCC plants mentioned above, only the Dutch Buggenum plant has been built without receiving any public subsidy.[43] Most of the other non-coal-fired IGCC plants that have

---

[42] The risk comparison is based on an evaluation of risk indicators such as legal problems, delays in procurement, change in project scope or plant site, shortage of skilled labor, labor disputes, redoing substandard work, equipment failure, contractor inefficiency, shortages of materials or fuels, uncontrollable events such as accidents, political turmoil, and natural disasters, and finally avenues for reducing risks. The cost figures provided in Table 5.2 are also factored into the evaluation. For a more detailed analysis see Nautilus Institute "Financing Clean Coal Technologies in China". Background paper for the ESENA Workshop. February 27-28, 1999.

[43] Mollot, Darren J. USDOE. "Clean Coal Technologies & Beyond: The Status of Prospects of Coal Power" Presented at the Johns Hopkins University School of Advanced International Studies (SAIS), April 28, 1999.

been constructed worldwide, without the receipt of public support, are fueled by cheap fuels, such as oil or petcoke, and/or rely on co-generation to improve the economics of electricity generation. Coal-fired IGCC technology has therefore not been able to reach a critical market penetration rate on either the world or Chinese market.

In conclusion, the application of the market penetration test confirms the results of the economic feasibility test that coal-fired IGCC technology is additional. The above analysis indicates that IGCC technology does not currently represent a viable option for China. However, it is possible that the favorable financing available through project- or market-based activities will change this picture and enable IGCC to compete on an equal footing with other energy alternatives. As a result, coal-fired IGCC technology should be included in the list of qualifying technologies in the technology matrix.

5.2.2.3 Temporal Considerations and Additionality. Although coal-fired IGCC technology is clearly not commercially viable at the moment, the economic disadvantages of the technology are not necessarily going to persist in the long term. IGCC cycle efficiency is expected to improve considerably over the next decade and costs are also expected to decline. The current level of efficiency is based on the technology available in the late 1980s and early 1990s. Since then, gasifier and turbine sizes have increased significantly and efficiencies have therefore improved. When the latest developments in gas turbines come into production, efficiency will rise to about 50 percent and costs will fall.[44] New hot gas cleanup (HGCU) technology also looks promising because it allows the syngas to be admitted to the combustion turbine without reheating.[45] Finally, experiments with air-blown coal gasification and advances based on studies of the optimum level of plant integration, particularly through integrated air systems are also likely to raise efficiency and decrease costs.

Likewise, it is possible that the policies in China affecting environmental regulation and availability of financing will change, thus impacting the economics and market penetration rate of IGCC systems. Environmental considerations are becoming more and more integrated into policy-making at the national and state levels thereby increasing the likelihood that, in the future, more stringent emissions controls will be imposed on the power sector. Furthermore, there are signs that China is increasing its involvement in the international economy. This may lead to a further deregulation of the electricity sector and a loosening of the restrictions on foreign direct investment and movement of capital. Such changes would have a positive impact on the economics of IGCC power plants and would lead to increased investment in the technology.

It is difficult to predict when or if all of these changes will take place. However, as they may eventually cause coal-fired IGCC technology to become economic in China, the changes would have the effect of making the technology non-additional. In such an event, coal-fired IGCC should be

---

[44] White, David. "Clean Coal Technology; Buggenum; A step towards commercialization of IGCC" *Modern Power System*. June 30, 1998.

[45] The Polk Power Station project in Polk County, Florida owned by Tampa Electric Company (TEC) accepts syngas at 900-1,000 F, eliminating the traditional step of cooling the gas to 100 F before removing the sulfur and then reheating the gas for the combustion turbine. *Power Engineering*, January 1997.

removed from China's modified technology matrix.

To prevent non-additional technologies from remaining on the list of qualifying technologies, the issue of additionality could be re-evaluated regularly, preferably every 5 years, for every technology on the list. Once it has been established that coal-fired IGCC technology is no longer additional and the technology has been removed from the list, new projects utilizing a coal-fired IGCC system will not be able to use the modified technology matrix approach to estimate the emission baseline. However, in case a group of project developers are convinced that their particular coal-fired IGCC project is additional and deserve credit under the market-based activity, they may instead use the project-specific approach to demonstrate the additionality of their project.

Meanwhile existing IGCC projects that gained approval while the technology still qualified as additional under the matrix approach will continue to earn credits. The fact that new coal-fired IGCC plants are no longer additional does not change the circumstances when the initial investment decision was made. Indeed, it may be the implementation of the project itself, which eventually improved the economics and the market penetration rate of the technology. Moreover, it is important to the project developers that they can recover the additional costs of undertaking a subeconomic project. In particular, investors in large-scale capital investment projects need a guarantee that they can generate credits for the entire economic life of the project in order to recover their fixed costs. Hence, the status of the project as additional will remain unchanged. However, the benchmark used to calculate the emission reductions of the project will be updated regularly to account for technological and other changes in the group of power plants that represent the benchmark. The procedures for updating the benchmark will be described in detail in section 5.3.5 of this chapter.

In sum, we established the additionality of the IGCC plant in China by applying the modified technology matrix approach. Following this approach, the additionality of the IGCC system was evaluated based on the qualities of the technology in general rather than on the circumstances of the individual project. To demonstrate the additionality of coal-fired IGCC technology in China, the economic feasibility and the market penetration test were applied. It was established that coal-fired IGCC technology is additional and should be added to the list of qualifying technologies. Consequently, our sample IGCC project will automatically qualify as a market-based project and will remain additional even though IGCC technology in general may in the future be removed from the China-specific modified technology matrix.

5.3 Establishing the Benchmark

Having considered the issue of additionality, the next step is to estimate the emissions baseline. Once again, from the standpoint of the *project developers*, estimating an emissions baseline using the modified technology matrix approach is a very simple and straightforward process. A stipulated benchmark will be provided for all participating countries and pre-qualifying technologies; the project developers need only identify the appropriate benchmark for their project, and use it as the basis for their emissions baseline.

In this section, we will therefore focus on the establishment of a benchmark for IGCC projects in China, rather than on the use of the benchmark for any particular project. The establishment of the benchmark must be based on a consideration of IGCC in general, rather than any particular project. This follows from the fact that the resulting benchmark must be applicable not only to the project described at the beginning of this chapter, but to all IGCC projects that might be undertaken in China. In fact, in order for the stipulated benchmark to be available to the project's developers, it must be established in advance of the project. Therefore, as was the case with our consideration of additionality in the preceding section, we must assume that the project described in Section 5.1 represents a *future* project, about which we have no knowledge for purposes of establishing the stipulated benchmark.

5.3.1 Qualitative Considerations

Although we must consider generalities rather than project specifics, the goal in establishing a benchmark under the modified technology matrix approach is similar to that in developing the emission baseline for the project specific approach. In both cases, we must try to ascertain, as closely as possible, the most likely alternative to the particular project or technology under consideration. With this ultimate goal in mind, the first question that must be addressed is as follows: will new IGCC power plants in China be used to replace existing capacity, or to meet new (or currently unmet) demand?

The Asian financial crisis has slowed Chinese economic growth. Due to the resulting short-term electricity-supply problem, China is shutting down some of the smaller and older inefficient coal plants and, as of 1999, has placed a two-year moratorium on the building of new power plants (three years for most coal-fired plants).[46]   However, continued rapid growth electricity demand growth (at an 8 percent annual rate) is expected over the longer term, as China continues its efforts to privatize state-owned industries, reduce the bloated government bureaucracy, attract foreign investment, and develop its infrastructure. Direct foreign investment in China has risen from $2.3 billion in 1987 to $43 billion in 1997; major investors include Japan, the United States, Singapore, and Taiwan. The importance of infrastructure improvement to the government is indicated by China's announcement of a 3-year, $750 billion infrastructure development program.

In response to China's expected long-term demand growth, the Energy Information Administration projects that electricity consumption will rise at an annual rate of 5.8 percent through 2020. This implies that China will require an additional 480 gigawatts of coal-fired capacity by 2020.

Given this situation, it is unlikely that a new IGCC power plant built in China will displace existing capacity. Thus be concluded that, in general, new IGCC power plants in China will serve the purpose of meeting new or currently unmet demand.

---

[46] U.S. Energy Information Administration (EIA), June 1999. http://www.eia.doe.gov/cabs/china.html

This being the case, we can proceed to the next question: how would this demand have been met in the absence of the typical IGCC project? Unlike the preceding question, the answer to this question is not clear cut. At the most general level, there are three possible answers:

1. The demand would in part have been met by small generators owned and operated by end users; the remainder would have been left unmet

2. A power plant using conventional subcritical technology, of similar size to the foregone IGCC plant, would have been built on the same site

3. New capacity would have been built to meet the demand, but this capacity would not have been located on the same site and might have consisted of a number of smaller plants, rather than a single plant of the same size.

The choice of a particular answer from among the above options has enormous implications for the stipulated benchmark. Consider, for example, the first answer. If the demand to be met by new IGCC power plants would, in the absence of the market-based activity, have been met only in part by small end-user generators, then the correct benchmark would lie between zero and the emissions rate of these small generators (most of which presumably utilize diesel fuel). If, on the other hand, the second answer was judged to be correct, then the benchmark would be based on the emissions of a conventional power plant probably utilizing the same fuel (coal) as the IGCC plant (in so far as there would have been no need to use IGCC technology if natural gas were readily available at the site). The benchmark implied by the third answer would also probably be based on the emissions rate of coal-fired capacity, if only because coal is the dominant fuel used by the Chinese electricity sector.

But which answer is in fact the most likely? First, it is important to draw a sharp distinction between developed and developing countries. In developed countries, electricity supply and demand are typically in balance, and the installation of new capacity keeps pace with demand growth. In these circumstances, if a decision is made to halt one project designed to meet new demand, it is likely that another project(s) will be implemented in its place. It is usually a given that the new demand will be met one way or another. Capital availability is typically not a constraint; if the demand exists, the capital required to build new capacity to meet that demand will generally be found.

The same is not true in the developing world. Capital availability acts a major constraint to capacity expansion. Demand is not always met, as evidenced by the significant electricity supply shortfalls in India and China before the Asian financial crisis. In these circumstances, it cannot be assumed that another capacity expansion project would automatically rise up to take the place of a forgone IGCC project. The possibility that the demand to have been met by the IGCC project will simply go unmet is a real one.

In short, capital availability, rather than demand, determines the amount of capacity that will be built. Therefore, the key question we must address is a financial question. In developed countries, a decision to forego a capacity expansion project will in effect "free up" (or leave unmet) a certain amount of excess demand; another project (or projects) will be undertaken and financed to fill this freed-up demand. In China and other developing countries, will a certain amount of capital likewise be "freed up" when a market-based IGCC project is forgone? If the answer to this question is "no," then we may conclude that the total amount of capital available to build new capacity will remain unaffected by the foregone IGCC project, and, given that capital availability determines the level of capacity expansion, there is no reason to believe that another project would be undertaken to replace the foregone project. If, on the other hand, forgoing the IGCC project *would* free up capital for another capacity expansion project(s), then such a replacement project(s) might presumably be undertaken.

To further our analysis of this issue, let us consider two possible alternatives to market-based IGCC project financing. In Alternative A, we assume that most of the financing is provided by the project's host country sponsors. The developed country sponsors provide a limited amount of additional financing in exchange for the emission reduction credits, but the host country sponsors retain all shares in the ownership of the project. What would happen if a decision were made to forego the project? In this particular situation, the capital that would have been provided by the host country sponsors would *probably* be freed up for another capacity expansion project. Possibly the loans and other financial instruments controlled by the host country sponsors might be lost once the IGCC project is foregone, but given that IGCC technology is not commercial at present it seems likely that the financing arranged for such a high-risk project could be retained for a project(s) utilizing conventional technology. Obviously, the host country sponsors could choose to invest the freed-up capital in projects other than capacity expansion projects, including projects in other sectors or countries. But again, if a capacity expansion project appears attractive despite its utilization of non-commercial, high risk technology, it seems plausible that the best alternative to such a project would be a capacity expansion project utilizing conventional technology. Obviously, the host country sponsors would lose the support of the developed country sponsors if they opted for a conventional power plant, because no emission reduction credits would be awarded for building such a plant. But because the capital costs associated with the construction of a conventional plant would generally be less than those required to build a comparable IGCC plant, and since the developed country sponsors are, in any event, providing only a small share of the required capital, it is likely that the host country sponsors could open a comparably sized conventional plant.

In Alternative B, we assume that the developed country-sponsors, rather than the host country sponsors, are providing most of the project financing. In this case, the developed country sponsors would presumably receive a significant share of ownership in the project, as well as the emission reduction credits, in exchange for their financial contribution. Given the large amount of capital required to open a power plant, and the fact that the developed country sponsors are providing most of the needed capital, it is likely that the ownership stake rather than the emission reduction credits are the prime motivator for developed country participation in the project. Hence the developed country sponsors are presumably interested primarily in exploiting perceived economic opportunities in the host country's electricity sector. The perceived market opportunities could presumably be met, at lower cost and lower risk, by opening a comparably sized conventional power plant. However, the added inducement of the emission reduction credits has caused the project sponsors to opt for an

IGCC plant rather than a conventional plant. Again it seems *plausible,* if not *probable,* that the next best investment alternative to the IGCC plant would be a conventional plant(s) in the host country.

In the case of financial arrangements lying somewhere between the extremes illustrated by Alternatives A and B, the argument, and the conclusions, would be similar.

Obviously, we cannot draw the conclusion that every foregone IGCC project would free up capital for conventional generating capacity; in at least some cases, it is possible that the demand to be met by the IGCC plant would simply go unmet in that plant's absence. In developing the technology matrix, we are, of necessity, constrained to consider entire categories of projects at once, rather than individual projects. Hence, our conclusions must be general, reflecting the likely typical situation, without taking into account the possible exceptions. This being said, it seems reasonable to conclude that, *in general,* the forgoing of an IGCC project in China would free up the capital needed to build a roughly equivalent amount of conventional capacity. The occasional exceptions to this general conclusion *will* reduce the accuracy of the emission reduction estimates derived using the modified technology matrix approach. But, as was discussed in Chapter 2, reduced accuracy *vis a vis* the project-specific approach must be accepted, along with reduced baseline development costs, as a characteristic of the modified technology matrix approach.

We may also conclude that the capacity likely to be built in lieu of the typical IGCC plant would take the form of a comparably sized conventional plant located on the same site. This conclusion follows from the arguments developed for Alternatives A and B above, and in particular from the fact that the investment capital to be freed up would remain under the control of the IGCC project sponsors. These sponsors, having selected the site for the IGCC project, would presumably favor the same site for the alternative plant. Furthermore, the alternative conventional plant would presumably use the same fuel (coal) as the IGCC plant (in so far as there would be no need to use IGCC technology if natural gas were readily available at the site). Again, these conclusions may be true in general, although they would not necessarily be true in every specific case.

Thus, *typically or in general,* the counterfactual for an IGCC project in China will be a coal-fired power plant utilizing conventional technology, of roughly comparable size to the IGCC plant and located on the same site.

5.3.2 Variable Versus Constant Benchmark

This counterfactual raises the possibility of the use of a *function*, rather than a constant, as the benchmark. The heat rate of a conventional coal-fired power plant (and hence, the plant's emission rate) is not a constant. Rather, it varies depending on such variables as the plant's design, capacity factor, and coal rank. *If* it is possible to assume that the counterfactual is in effect a "shadow" version of the IGCC plant, located on the same site and subject to the same variation in parameters, then a function relating these parameters to the shadow plant's emissions could perhaps be used to estimate an annual, varying emissions baseline. In other words, rather than setting the benchmark (B) equal to a constant value, we might instead define the benchmark as a function of parameters such as coal quality (Q) and the capacity factor (CF); i.e.

$$B_j = f(Q_j, CF_j)$$

The values for Q and CF in any given time period j would be set equal to the actual average coal quality and capacity factor experienced at the IGCC plant during the same time period. The use of such a function is made possible by our assumption that the counterfactual is a coal-fired plant located on the same site as the IGCC plant, and of the same size. Obviously, if the counterfactual were instead assumed to be a number of smaller plants scattered throughout the country, the values of Q_j and CF_j for the IGCC plant could not be treated as equivalent to the values of these variables at the counterfactual plants.

A functional approach to benchmark development might prove more accurate than a constant benchmark, because it would capture the impact of key variables on the counterfactual as well as the IGCC plant. The heat rates of conventional coal-fired plants can vary significantly depending on these variables; hence, a *reliable* procedure for capturing these variations should improve the accuracy of the resulting baseline emission estimates.

However, there are some drawbacks and potentially serious problems with a functional approach. First and foremost, the data required to support such an approach is much more extensive than that required to develop a constant benchmark. Rather than simply obtaining average heat rates for China's conventional coal-fired power plants, it would be necessary to collect data on the key determinants of heat rate as well heat rate data itself. Even by developing country standards, China's energy data is unreliable and difficult to access. As an example of the extent of the problem, it has been estimated that China's national coal production and consumption statistics may be under-reported by as much as 200 million tons. In this environment, it is not clear that reliable information on variables, such as coal quality, would be forthcoming without a major investment in new data surveys, and perhaps, new regulations mandating the reporting of the needed data. Furthermore, in order to update the benchmark function with new data, it would be necessary to repeat the required surveys on a periodic (probably annual) basis.

The additional expenditures in time, money, and resources required to meet the extensive data requirements *might* be justifiable if it could be demonstrated that they led to a significant improvement in baseline accuracy. However, a functional approach will not necessarily yield such an improvement. This follows from the fact that the counterfactual will never perfectly "shadow" the IGCC project. For one, even assuming that both the IGCC and counterfactual plants generate the same quantity of electricity, they will not consume the same amount of fuel given the differences in their heat rates. Specifically, the counterfactual plant will consume more coal, in a given time period, than the IGCC plant. This in turn means that the coal quality variations experienced by the IGCC plant in any given time period j will differ from the variations that would have been experienced by the counterfactual plant during the same time period. Furthermore, the capacity factors (and hence the net generation)

106

of the two plants cannot be assumed to be the same in any given time period. Because the technology employed by the IGCC plant is new, experimental, and more complex than that utilized by a conventional coal-fired plant, it is likely that it would experience more scheduled and unscheduled outages than the counterfactual plant. Finally, the assumption that the counterfactual plant would have the same capacity as the IGCC plant *is* just an assumption; in reality, it is likely that there would be some differences.

For these reasons, the values of Q_j and CF_j for the IGCC plant will provide at best only a rough approximation of the corresponding values for the counterfactual plant. This being the case, the value of the benchmark (B_j) yielded by the above function will capture variations in heat rate determinants only in a rough, imprecise manner. In short, it is not clear that the utilization of a functional approach to benchmark determination would represent a significant improvement over the use of a constant. Hence, given the difficulties associated with gathering the data required to specify the function, and the reliability questions surrounding such data, we recommend that a constant benchmark, rather than a variable benchmark, be utilized. It should be emphasized that this recommendation is specific to China; the issue of variable versus constant benchmarks should be revisited in the case of developing countries with relatively reliable, accessible data.

Also, our rejection of the functional approach to benchmark development should not be construed as precluding the *updating* of the benchmark on a periodic basis. In fact, a benchmark that varies with *time*, to capture changes at China's conventional power plants, should significantly enhance accuracy. As we shall see below, such a time-dependent benchmark *will* capture large-scale trends in coal quality and capacity utilization, although site-specific variations will not be captured. Thus, although we are recommending the use of a constant rather than a function for the benchmark *at any particular point in time*, the constant will itself change over time.

5.3.3 Quantifying the Benchmark

Having decided upon a constant benchmark, we may now consider the metric that will be used as the benchmark. Discussions of the benchmarking approach have typically proposed that an emissions rate be used as the benchmark value. In the power sector, this emissions rate would be expressed in pounds of carbon dioxide per kilowatt-hour. However, it is believed that a benchmark based on heat rate (expressed in Btus per kilowatt-hour) would represent an improvement over an emissions-rate benchmark. A power plant's emission rate will vary slightly depending on the type of coal burned. In the United States, carbon dioxide emissions vary from about 205 pounds per million Btu for bituminous coal up to 213 pounds per million Btu for subbituminous coal and 227 pounds per million Btu for anthracite. By basing the benchmark on heat rate rather than a per kilowatt-hour emissions rate, it will be possible to control for this variation. Specifically, by multiplying the heat rate benchmark by the appropriate emissions coefficient for the type of coal used by a given IGCC project, the benchmark can be converted into an emissions rate appropriate for that project; that is:

$$ER_x = (BHR)(EC_y)/(1,000,000 \text{ Btus/mmBtu})$$

Where:

ER_x = The benchmark emissions rate for a given IGCC project x, in pounds of carbon dioxide per kilowatt-hour

BHR = The benchmark heat rate for all IGCC projects in China, in Btus per kilowatt-hour

EC_y = The emissions coefficient for coal of rank y in pounds carbon dioxide per million Btus, where y is the rank of coal consumed by IGCC project x

The emissions rate ER_x computed using the above equation can be converted into the baseline emission estimate for any given IGCC project x, in year j, simply by multiplying by the project's total net generation in year j:

$$BE_{xj} = (ER_x)(G_{xj})$$

Where:

BE_{xj} = baseline emissions (in pounds carbon dioxide) for IGCC project x in year j

G_{xj} = total net generation of project x in year j (kilowatt-hours)

The actual emissions of project x in year j would be subtracted from the baseline emissions computed above to yield the project's estimated emission reductions. Note that the above approach implicitly assumes that the counterfactual power plant would generate the same amount of electricity as the IGCC plant. This assumption is a necessary simplification.

It remains only to specify BHR, the benchmark heat rate, to complete the above equations. As we have already established, the assumed counterfactual for the standard IGCC project in China is a conventional coal-fired power plant, with the same capacity, and located on the same site as, the project. Furthermore, because IGCC projects in China will invariably involve the opening of new capacity (there is no existing IGCC capacity in China), the counterfactual should likewise represent new capacity. Based on these considerations, we propose that the benchmark heat rate represent an average heat rate for recently-constructed coal-fired units in China. By limiting the average to recently opened capacity, we can better assure that the benchmark will reflect the typical heat rate for conventional plants likely to be opened in the near-term future; older plants, many of which may be less efficient as a result of the wear and tear of age, will be eliminated from the computation of the heat rate average.

The age cutoff for inclusion of plants in the benchmark computation must depend, to some extent, on data availability. Five years seems a reasonable age cutoff, *if* heat rate data for a sufficiently large sample of power plants built in the last five years can be obtained. Separate benchmarks should be developed for generating units of different size (capacity) ranges. Three different capacity ranges should suffice.

In summary, three separate benchmarks should be developed representing the average heat rate for three different size ranges of conventional coal-fired generating units, built in China during the last 5 years. If, however, data are only available for a few such power plants, it may be necessary to extend the age cutoff beyond five years.

5.3.4 Regional Considerations

In a country the size of China, there are significant differences in such factors as coal quality, water availability, financial resources, plant site availability, the size of the electricity supply shortfall, etc., all of which may affect the design and operation (and therefore the heat rate) of conventional power plants. Therefore, at least from a theoretical perspective there are advantages to be gained, in terms of accuracy, by dividing the country into regions and developing separate benchmarks for each region. However, the decision to utilize regional benchmarks, and the definition of appropriate regions, must be based on empirical analysis rather than *a priori* judgement. First, it is necessary to analyze the available heat rate data to determine whether or not, and where, significant regional differences exist. Assuming such differences are found, regions can be defined that group the existing plant population to accentuate, or at least reveal, these differences. However, consideration must also be given to the size of the resulting regional sub-populations. Specifically, all regions must be of a size sufficient to ensure an adequate sample of plants upon which to base the benchmark. Also, any data availability constraints must be taken into account when sizing and defining the regions. In short, the issue of whether or not to regionalize the Chinese benchmarks is worthy of further analysis, but this analysis must await the collection of the required data.

5.3.5 Temporal Considerations

Clearly, it will be necessary to re-estimate the average heat rate on a periodic basis to reflect changes or improvements in the operating efficiencies of new coal-fired power plants. It is believed that re-estimation once every 5 years will be sufficient to keep the benchmark up to date, since the average heat rate of new conventional steam plants tends to be fairly stable over time.

The most up-to-date benchmark available should, of course, always be used for new IGCC projects. However, although updates will be required for existing market-based IGCC projects, the methodology to be employed for updating the benchmarks for ongoing projects will differ from that used for new projects. Here, it is important to recall our assumption that, but for the project, a conventional coal-fired power plant would have been built on the project location. Steam turbine power plants are long-lived facilities. It is not uncommon to find power plants that have been in operation for over 50 years in the United States, and in China the impetus for extending the life of existing plants is even greater given the electricity supply shortfalls. If we were to update the benchmark for existing power plants every five years, based on the heat rate of recently-constructed plants, then we would in effect be assuming that the counterfactual power plant would be shut down and replaced with a new power plant every five years. Such an assumption is clearly unrealistic.

However, the heat rate of the counterfactual power plant would clearly change over time. Heat rates naturally tend to increase with power plant age, as a result of wear and tear on the equipment. To combat this problem, plant operators replace worn parts and equipment on a regular, and fairly frequent, basis. Furthermore, on a less frequent basis, major overhauls are undertaken to improve plant availability and efficiency. Repowering projects, which may involve the retrofitting of new technologies to existing plants, may also be undertaken on occasion. Also, as power plants reach the end of their design lives, life extension projects may be undertaken. All of these projects usually have the effect of improving efficiency, just as the passage of time has the effect of reducing efficiency.

How do we take into account the effect of time, and of efficiency improvement efforts, on the benchmark heat rates for ongoing IGCC projects? In developing the benchmark for *new* projects, we will in effect be identifying a group of actual power plants that will be taken as representative of the counterfactual plant. The average heat rate for this group of plants will be used as the benchmark for the first five years of each project's existence. Because this group of plants has been selected as being representative of the counterfactual, we can trace developments at these plants over time, and update the benchmark every five years using the average heat rate for the same group. In effect, the group of power plants originally selected as the basis for the benchmark will continue to serve as the benchmark throughout each project's life. In this way, technology improvements that affect *existing* power plants, such as the retrofitting of new boiler designs to such plants, will be captured as these improvements penetrate the benchmark group.

In effect, we are suggesting the addition of a third dimension to the technology matrix illustrated in Table 2.2 (Chapter 2): time. At first, a group of benchmark power plants will be selected from the population; all market-based IGCC projects initiated in the subsequent five years will use that group of plants as their benchmark throughout their lives. After 5 years has elapsed, a new benchmark group will be selected; all projects implemented in the next five years will use this new group of plants as their benchmark. Hence, for any given qualifying technology and country, a series of benchmarks will be in use at any given time for new and ongoing projects; the specific benchmark used by a particular project will be determined by that project's start date. Table 5.3 provides an illustrative example of the technology matrix, modified to incorporate a time dimension.

Under this approach, the benchmark heat rate will not be a constant, but will rather vary with time. Furthermore, although the determinant of the benchmark will be time rather than coal quality or capacity utilization, the time variable will in fact capture *large-scale* trends in these other variables. For example, if, in general, the quality of Chinese coal deteriorates through depletion of the higher-quality reserves, this deterioration will have an adverse impact on the heat rates of China's power plants, including those plants used as benchmarks. As a result the benchmark heat rates will rise.

110

Similarly, if the current gap between electricity supply and demand is reduced and eventually eliminated, this may affect the capacity factor at China's existing plants, which should in turn have an impact on the benchmark heat rates. Thus, although site-specific volatility in the determinants of heat rate will not be reflected in the benchmark, the impact of widespread, long-term trends in these determinants will be captured. Given the inherent imprecision in the use of benchmarks and the impossibility of synchronizing the counterfactual with the site-specific changes experienced by the project, the use of time to indirectly capture large-scale trends in heat rates seems a reasonable alternative to the functional approach described above.

It should be emphasized that, using this approach, the benchmark heat rate may decrease or *increase* over time, depending in part upon the vigilance of China's efforts to maintain existing power plants. If, in general, China fails to adequately maintain its power plants, this failure will lead to a deterioration in the plants' heat rates, and this deterioration should be reflected in the benchmark. If, on the other hand, China maintains its heat rates and even improves upon them by introducing new technologies into its existing plants, this improvement should likewise be captured in the benchmark. The goal, in all cases, is to trace the time path most likely to be followed by the project counterfactual, and in so doing to maintain a set of benchmarks that provide a realistic, credible reflection of developments in the Chinese power sector.

Finally, it should be noted that in some cases a project may outlive the group of power plants upon which its benchmark is based. If, at some point in time, over 50 percent of the capacity forming a particular benchmark group has been retired, then it would be reasonable to conclude that the project counterfactual represented by this group has been retired. At this point in time, it would be reasonable to switch the benchmark for any projects still in existence from this old group of plants to the group used to define the most up-to-date benchmark (i.e., the group of plants opened within the past five years). The assumption underlying this benchmark switch is that the original counterfactual power plant has been retired, and replaced with a new counterfactual plant.

5.4 Summary

In this chapter we developed the baseline for a hypothetical coal-fired Integrated Gasification Combined Cycle (IGCC) project in China. Because IGCC technology is expected to qualify as additional under the modified technology matrix approach, we selected this approach for application to the project. Furthermore, because the application of the technology matrix to any particular project is a relatively trivial exercise, we instead focused on development of the matrix for IGCC technology in China. We began the development process by establishing the additionality of IGCC technology. Specifically, it was established that, given the present capital and total costs of this technology, coal-fired IGCC cannot compete economically with conventional power generation technologies, either in China or in the world as a whole. This conclusion was backed up by a consideration of market penetration. No IGCC projects have been developed in China to date, and of the 5 commercial-scale IGCC projects in existence throughout the world, only one is operating without public subsidies.

111

Table 5.3. Example of a Portion of the Technology Matrix with the Initial Benchmarks and the First Two Five-Year Updates

Countries / Qualifying Technologies		India Year 1	India Year 6	India Year 11	China Year 1	China Year 6	China Year 11	Argentina Year 1	Argentina Year 6	Argentina Year 11
Coal-Fired IGCC	BMG a	B_a	B_{a+5}	B_{a+10}	B_a	B_{a+5}	B_{a+10}	B_a	B_{a+5}	B_{a+10}
	BMG b	---	B_b	B_{b+5}	---	B_b	B_{b+5}	---	B_b	B_{b+5}
	BMG c	---	---	B_c	---	---	B_c	---	---	B_c
Solid Oxide Fuel Cells	BMG a	B_a	B_{a+5}	B_{a+10}	B_a	B_{a+5}	B_{a+10}	B_a	B_{a+5}	B_{a+10}
	BMG b	---	B_b	B_{b+5}	---	B_b	B_{b+5}	---	B_b	B_{b+5}
	BMG c	---	---	B_c	---	---	B_c	---	---	B_c
Phosphoric Acid Fuel Cells	BMG a	B_a	B_{a+5}	B_{a+10}	B_a	B_{a+5}	B_{a+10}	B_a	B_{a+5}	B_{a+10}
	BMG b	---	(shaded)	(shaded)	---	B_b	B_{b+5}	---	(shaded)	(shaded)
	BMG c	---	---	(shaded)	---	---	(shaded)	---	---	(shaded)
Molten Carbonate Fuel Cells	BMG a	B_a	B_{a+5}	B_{a+10}	B_a	B_{a+5}	B_{a+10}	B_a	B_{a+5}	B_{a+10}
	BMG b	---	B_b	B_{b+5}	---	B_b	B_{b+5}	---	B_b	B_{b+5}
	BMG c	---	---	B_c	---	---	B_c	---	---	B_c
Proton Exchange Membrane Fuel Cells	BMG a	B_a	B_{a+5}	B_{a+10}	B_a	B_{a+5}	B_{a+10}	B_a	B_{a+5}	B_{a+10}
	BMG b	---	B_b	B_{b+5}	---	B_b	B_{b+5}	---	B_b	B_{b+5}
	BMG c	---	---	B_c	---	---	B_c	---	---	B_c
Photovoltaics	BMG a	B_a	B_{a+5}	B_{a+10}	B_a	B_{a+5}	B_{a+10}	B_a	B_{a+5}	B_{a+10}
	BMG b	---	B_b	B_{b+5}	---	B_b	B_{b+5}	---	B_b	B_{b+5}
	BMG c	---	---	B_c	---	---	B_c	---	---	B_c
Pressurized Fluidized Bed Combustion	BMG a	B_a	B_{a+5}	B_{a+10}	B_a	B_{a+5}	B_{a+10}	B_a	B_{a+5}	B_{a+10}
	BMG b	---	B_b	B_{b+5}	---	B_b	B_{b+5}	---	B_b	B_{b+5}
	BMG c	---	---	B_c	---	---	B_c	---	---	B_c

Notes: 1) BMG = Benchmark group; a group of generating units selected as representative of the counterfactual.

2) B_x = Estimate of benchmark emissions or heat rate for benchmark group x

3) B_{x+5} = Five year update of the original benchmark for benchmark group x (B_x)

4) B_{x+10} = Ten year update of the benchmark

5) Shaded areas represent technology/country combinations that do not qualify as additional

After establishing the additionality of IGCC technology, we considered the issue of defining an appropriate benchmark for this technology. We concluded that the benchmark should represent the average heat rate of conventional coal-fired power plants installed in China in the last five years. It was further decided that a benchmark for *new* projects should be developed once every five years, while the benchmark for *existing* projects should be updated to reflect changes in efficiency for the same group of power plants used as the basis for the original benchmark.

We will return to the China IGCC project in Chapter 7, where we will subject the three case studies to a critical assessment.

6. FUEL CELLS IN ARGENTINA

6.1 Project Description

6.1.1 Background [47]

The Argentinian Fuel Cell project, like the Chinese IGCC project, is hypothetical in nature. At present, there are no operating fuel cell generators in Argentina, nor are there any demonstration projects. However, the Argentinian government has a strong interest in the development of fuel cell technology. At the government's request, NETL has given a number of presentations on fuel cells in Argentina. Furthermore, Energy Research Corporation (ERC), a U.S. company, is actively pursuing a demonstration project in Argentina. ERC is proposing a molten carbonate fuel cell design.

Compared to India and China, Argentina's electric power system is well developed. The electricity grid is relatively extensive and reliable. Argentina has excess generating capacity and sells power to neighboring countries. At present, thermal power plants, relying primarily on natural gas (and, to a lesser extent, oil) comprise 60 to 70 percent of Argentina's capacity. Hydropower comprises 30 percent of capacity, and nuclear power most of the remainder.

Argentina is experiencing an annual two to five percent growth in electricity demand, and much of this growth is occurring in rural areas. Expansion and extension of the power grid has not kept up with this rural demand. Currently, an estimated 500 to 1000 rural villages rely on diesel generators to meet their power needs. However, due to wide daily fluctuations in load, these generators often operate at very low efficiencies. Typically, the generators achieve their maximum efficiency when they operate at about 80 percent of capacity; however, the actual demand load fluctuates from nearly zero during the day to peak load conditions in the evening. The Argentine government is interested in fuel cells as a potential high-efficiency replacement for the diesel generators.

There is an issue of fuel availability in these areas that has not yet been fully addressed. However, Argentina does have a natural gas pipeline network, and in fact is an exporter of natural gas to Chile. A number of gas transmission lines pass through rural areas. Only a fraction of Argentina's rural villages are sited near existing transmission lines, so potential fuel cell applications may be quite limited. However, for such villages fuel cells operating on natural gas are a real option.

6.1.2 The Project

The project will involve the use of solid oxide fuel cells (SOFCs) as off-grid power generators for a rural village. The local utility sponsoring the project has considered a number of options for bringing electricity to this and other villages in its service territory. One option would involve the construction of new power lines. However, owing to the villages' remote location, and the long distance such a

---

[47]A major part of the information presented in this subsection was obtained via personal communication with Maria Reidpath of NETL.

114

line would have to traverse, this would be the most capital-intensive of all the options considered. A second possibility would be to continue to rely on diesel generators. This option would be economically preferable to the transmission line option. However, instead of diesel generators, the utility has opted for SOFC fuel cells, because, with rapid advances in fuel cell technology expected, this technology may ultimately prove to be the most efficient, cost-effective means of electrifying rural villages in the coming years. To limit the financial risk involved, the utility is setting the project up as a small demonstration project. Initially, the number of fuel cells will be limited to one 250-kW unit. If the demonstration proves successful, the project will be expanded to include additional units. If, on the other hand, the fuel cells fail to meet the utility's performance requirements, the utility will re-evaluate its options, and may choose instead to deploy (or redeploy) diesel generators. Because the project has been limited to a small demonstration, the financial loss to the utility in the event of failure will be small.

In addition, the financial risk will be further minimized by a subsidy provided by a U.S. utility. The U.S. utility has agreed to provide 25 percent of the required investment, in exchange for the emission reduction credits to be awarded under the market- or project-based activity. The U.S. utility is interested in SOFCs for potential applications in its own service territory, and is looking upon the demonstration as a means of gaining experience with the technology while minimizing its own risk. If the unit operates as anticipated, the U.S. sponsor expects to recover its costs in the form of emission reduction credits. Hence, the project's cost-sharing aspects, in combination with the potential for earning credits, reduce the risks to both parties to an acceptable level.

It should be noted that the Argentine utility's interest in fuel cells is being driven ultimately by their potential applicability to other rural villages within the utility's service territory. Hence, the benefits to be gained if the project proves successful are extensive, compared to the project's risks.

The project sponsors have selected solid oxide fuel cells for a number of reasons. For one, SOFCs have the potential to span market competitive applications from residential loads as low as two kW to distributed generation units of 25 MW. This flexibility is considered a major advantage of SOFCs by the two utility partners, because both have a wide range of potential fuel cell applications (beyond rural applications). Furthermore, SOFCs operate at high temperatures (850-1000^{0}C), and the waste heat can be used to drive small gas turbines. Microturbines that could be combined with SOFCs to achieve overall system efficiencies of at least 60 percent, and perhaps more than 70 percent, are currently under development. Even when not combined with gas turbines, SOFCs can achieve simple system efficiencies of approximately 50 percent. The project will be limited to a demonstration of SOFCs, but both utilities expect to pursue future projects combining SOFCs with gas turbines once the fuel cell technology has been successfully demonstrated.

Fuel cells operate by converting the chemical energy in a hydrogen-rich fuel into electricity and heat, without combusting the fuel. In a conventional phosphoric acid fuel cell (PAFC), the required fuel is produced by a separate fuel processor through steam reforming of a fossil fuel such as natural gas. Another advantage of SOFCs is that, due to their high operating temperatures, an internal reformer that uses heat from the fuel cell--along with recycled steam and a catalyst -- can be incorporated into the fuel cell design. The demonstration project will use natural gas as the fuel; gas is readily available

115

in the project area due to the nearby location of a gas transmission line. A small gas pressure reduction station will be installed near the village to provide low-pressure natural gas to the fuel cells.

Because SOFCs use a solid ceramic electrolyte rather than the liquid electrolyte used by PAFCs, electrolyte containment problems are eliminated and material corrosion is reduced. Finally, the project sponsors are attracted to SOFC technology because it has the potential for achieving cost targets of $1000 to $1200 per kW if it is combined with a natural gas turbine; $1500 per kW is the cost target where it is believed that SOFC will become commercially competitive for distributed generation applications.

6.2 Emission Baseline Development

6.2.1 Evaluation of Baseline Options

For this project, as for the previous two projects, choosing between the project-specific and modified technology matrix approach is simple and straightforward. Referring to Table 3.1 in Chapter 3, there is little question that this project falls within the Type 1 project category, i.e., new capacity projects that use advance qualifying technologies. Phosphoric acid fuel cells are beginning to penetrate the market in some areas, and now or in the near future might no longer qualify for inclusion in the modified technology matrix. However, solid oxide fuel cells remain in the experimental stage of development, and projects utilizing SOFCs should automatically qualify as additional under the modified technology matrix approach.

As in the case of the China IGCC project, the modified technology matrix approach will be used. Once again, we will focus on the *development* of the modified technology matrix for fuel cell technology in Argentina rather than its *application* to the particular project described in the above section. Applying the modified technology matrix to any particular project should be a relatively trivial exercise. The project developers need only demonstrate that they are in fact using a technology included in the matrix to prove that their project is additional, and the project's baseline can be readily derived from the emissions benchmark provided in the matrix. On the other hand, developing the matrix for any particular country and technology is not a trivial exercise.

In the remainder of this chapter, we will develop the modified technology matrix for SOFC technology in Argentina, by (1) demonstrating the additionality of this technology and (2) developing an appropriate emissions benchmark for the technology. It is necessary to treat each of the four fuel cell technologies--phosphoric acid, molten carbonate, proton exchange membrane, and solid oxide--separately within the modified technology matrix, for these four technologies are at different stages of development. As mentioned above, phosphoric acid fuel cells are in the initial stages of commercialization and perhaps may not qualify for inclusion in the modified technology matrix; the other three technologies are non-commercial and should meet the qualification criteria. However, because our example project involves only solid oxide fuel cell technology, we will focus exclusively on SOFCs.

116

Finally, as was the case for the China IGCC project, the modified technology matrix for SOFCs in Argentina must be developed without any reference to the specific project described above. The modified technology matrix is a *generic* approach designed to be applicable to *all* projects using a particular technology, it must therefore be developed based on a consideration of the typical project rather than any particular project. Furthermore, the modified technology matrix will be developed in advance of the particular projects that will utilize the matrix. We must assume, for matrix development purposes, that the project described above is a *future* project, and that the matrix developers therefore cannot have any knowledge of this project. Therefore, no further reference will be made to our SOFC project, although we will refer to the general background information presented at the beginning of this chapter.

6.2.2 Additionality

The next step in the baseline development process entails demonstrating the additionality of SOFC technology. In the previous section, it was established that the modified technology matrix should be used for developing the emission baseline. Therefore, to apply the additionality test we are going to examine the economic feasibility and the market penetration rate of SOFC technology in general rather than evaluating the conditions of the individual fuel cell project in Argentina. We follow this approach in order to demonstrate the issues involved in applying the additionality test under the modified technology matrix approach. Normally, a project developer would not have to go through this step, as the list of qualifying technologies would already have been developed by the host country and the administrators of the market- or project-based activity. However, as the modified technology matrix has not yet been developed, this section will proceed to evaluate the additionality of SOFC technology for the purpose of adding it to Argentina's technology matrix.

In the following subsections, we will demonstrate the additionality of SOFC by evaluating the economic feasibility and the market penetration of the technology. SOFC is found to be unable to compete economically with existing technologies and has failed to reach even a minimal market penetration rate in Argentina. Therefore, the technology qualifies as additional.

6.2.2.1 Economic Feasibility of SOFC. The first step in determining additionality involves evaluating the economics of SOFC technology. Thus, the cost of installing a SOFC generator should be compared to the cost of building a similar-sized power plant using alternate, market-proven technologies. The SOFC technology will then qualify as additional if it is proven that it will only be economic if it receives the favorable financing available through participation in the market- or project-based activity.

Several demonstration projects using SOFC technology have already been implemented.[48] Westinghouse Electric Corporation's Science & Technology Center is the current leader in terms of the number and size of test units implemented. Among other activities, Westinghouse is installing an experimental 200-kW system in a cogeneration application in the Netherlands in cooperation with a

---

[48] Moore, Taylor. "Market Potential High for Fuel Cells" *EPRI Journal*, May/June 1997.

consortium of Dutch and Danish utilities. Moreover, Westinghouse has built a $132 million pilot manufacturing facility, capable of producing about 4 MW of SOFCs a year. The company is also planning the construction of a commercial SOFC manufacturing facility to target mainly multi-megawatt systems in the range of 30-100 MW. These systems would combine fuel cells with a gas turbine (SOFC-GT) in packaged units, thus increasing efficiency to 60-70 percent. Other companies developing SOFC technologies include Ztek Corporation, which is working together with the Electric Power Research Institute (EPRI) and the Tennessee Valley Authority (TVA) to install experimental systems, SOFCo, Technology Management Incorporated, and AlliedSignal Corporation. Moreover, at least seven companies in Japan, eight in Europe, and one in Australia are working on developing SOFCs at the moment.

Even though the application of SOFC technology is growing in the distributed electricity market, it is still not considered a proven or mature technology. In addition, the operational performance of SOFC systems is still being tested, although results achieved so far promise acceptable component life characteristics. Moreover, the cost of SOFC technology has still not reached a level where it is competitive on the electricity market. SOFC technology is expected to be able to compete in high-cost, distributed electricity generation markets once it falls below $1,500 per kW. However, the price of fuel cells currently falls in the range of $3,000-$5,000 per kW. As a third generation technology, SOFC systems fall in the high end of this spectrum.[49]

A comparison with other energy sources highlights the high costs of SOFC technology. Referring to Table 5.1 in Chapter 5, we find that the capital cost of a conventional pulverized coal (PC) plant with flue gas desulfurization (FGD) controls ranges between $1000 to $1200 per kW while the cost of a natural gas-fired combined cycle plant lies between $400 to $500 per kW. Based on these numbers, SOFC technology is clearly not economic at the moment, and particularly not in Argentina. Energy costs are low in Argentina as a result of the country's recent liberalization of the energy market and the heavy reliance on low-cost energy sources, such as natural gas and hydropower.[50] In addition, interests rates are very high in Argentina, leading to a general preference for low capital cost inves
Argentina.

To summarize, the economic feasibility test indicates that SOFC systems are currently not commercial in Argentina and in rest of the world. Unless favorable financing is provided through the market- or project-based activity, SOFC technology is not likely to be introduced on a wide scale in Argentina in the near future. Therefore, we conclude that SOFC technology is additional and should be added to the modified technology matrix for Argentina.

---

[49] "Fuel Cells 2000: Frequently Asked Questions about Fuel Cells."
http://www.fuelcells.org/fuel/fcfaqs.html
[50] U.S. Energy Information Administration. Argentina Country Analysis Brief.
http://www.eia.doe.gov/emeu/cabs/argentina.html
[51] "Air Pollution Control Measuring Equipment: Argentina." Market Research Reports: Industry Sector Analysis. US Department of Commerce, International Trade Administration. May, 1997.

6.2.2.2 Market Penetration of SOFC in Argentina. Another method for establishing additionality under the modified technology matrix approach involves evaluating the market penetration rate of the particular technology. If it can be established that the technology in question has not reached even a minimum threshold of market penetration without benefitting from public subsidies or other favorable treatment, it should qualify as additional.

As in the case of IGCC technology in China, the question of establishing the market penetration rate of SOFC technology in Argentina is fairly straightforward because no fuel cells of any type have been installed in Argentina. In general, very few clean technologies have reached the Argentine market because environmental regulation and enforcement, in particular, is very weak. As a result, companies face few incentives to experiment with cleaner but more expensive and high-risk technologies.

Solid oxide fuel cells have not reached a wide market penetration on a global basis either. As described in the previous section, several companies are involved with developing SOFCs and the technology has been installed in numerous test sites and research facilities. However, these activities have mainly come about through public research support and other incentives. For example, the Department of Energy is heavily involved in the research and development of fuel cell technologies. The governments of Canada, Japan, and Germany are also promoting fuel cell development via tax credits, low-interest loans, and grants to support early purchases and lower costs. The strong public involvement in the development of SOFCs suggests that this is a technology which has not yet reached a fully commercial stage.

In conclusion, SOFCs have not yet penetrated the Argentine market, providing another indicator that the technology is additional and should qualify for inclusion in the modified technology matrix.

6.2.2.3 Temporal Considerations and Additionality. In the previous two sections, it was established that SOFC technology is not economically viable at the moment. However, the economic disadvantages of SOFC technology may not persist in the long term. The economies of scale of fuel cell technology are significant. Current plans to build commercial scale manufacturing plants, such as the one under construction by Westinghouse, are expected to bring down the cost of SOFC systems dramatically. Indeed, Westinghouse expects to reach a capital cost level of $1,300 per kW around 2003. In addition, it is possible that the regulatory environment in Argentina will change, creating an impetus for implementing cleaner energy technologies, such as fuel cells.

To account for these possible changes in the economics and the market penetration rate of SOFC technology, the additionality of SOFCs should be reevaluated regularly, preferably every five years. If one of these reevaluations demonstrate that SOFCs are no longer additional, the technology should be removed from the modified technology matrix. Developers of new projects based on SOFC technology would then be required to apply the project-specific approach to baseline development in order to demonstrate additionality. Meanwhile, projects that were approved while the SOFC technology was still additional should continue to receive credit. As noted in the analysis of the Chinese IGCC project, investors need a guarantee that they will be able to recover their initial costs. However the baseline against which the credits of these projects are determined, should be updated regularly to account for technological and other changes in the benchmark group representing the counterfactual. The procedures for updating this benchmark will be described in detail in section 6.3.4 of this chapter.

To summarize, we selected the modified technology matrix approach to develop the emission baseline for solid oxide fuel cells. Hence, additionality was demonstrated by analyzing SOFC technology in general rather than examining the specific fuel cell project in Argentina. The economic feasibility and market penetration tests were applied to demonstrate the additionality of SOFCs and it was determined that the technology should be added to the modified technology matrix. Consequently, the fuel cell project in Argentina will automatically qualify as a market-based project and will remain additional even if SOFCs become commercial and are removed from the list of qualifying technologies.

6.3 Establishing the Benchmark

6.3.1 The Most Likely Alternative

Having addressed the additionality issue, we may now turn to the development of the emissions benchmark for SOFC technology in Argentina. We must begin by considering the following basic question: what is the most likely alternative to the qualifying technology? Fuel cells have a wide variety of stationary power applications. For example, they can be used to provide on-site power at remote locations not connected to the grid, they can be used for distributed generation, or they can be used to provide premium power to end users with critical loads (such as hospitals) or special power requirements (like data processing centers). As a result of their high operating temperatures, SOFC produce waste heat that can be recovered for cogeneration applications. Given these myriad applications, the potential project alternatives could range anywhere from small off-grid diesel generators to major electric transmission capacity improvement projects.

Establishing a single most likely project alternative, in the face of this wide array of potential applications, could prove a daunting task were it not for the knowledge NETL has obtained on the Argentine government's goals and objectives regarding fuel cells. Recall, from the introduction to this chapter, that Argentina is currently experiencing electricity demand growth of two to five percent per year, and that much of this growth is occurring in rural areas. Recall, further, that the Argentine government is interested in fuel cells particularly for some off-grid applications, to meet the growing rural demand. Finally, because Argentina currently relies heavily on diesel generators to meet rural demand, we may conclude that the most common or typical alternative to stationary fuel cell projects is likely to be diesel generators.

Therefore, we will base the emissions benchmark on the emissions characteristics of diesel generators in Argentina. Note, however, that another possible approach to this situation would be to develop separate benchmarks for different fuel cell applications. For example, one benchmark might be developed for off-grid fuel cells, another for cogeneration applications, etc. One of the advantages of the modified technology matrix is its flexibility. Any particular technology to be included in the matrix can be disaggregated based on technical specifications, applications, or any other set of criteria depending on the circumstances or goals; in this way the benchmarks can be developed and refined to fit as narrow (or as broad) a set of projects as desired.

However, because it appears that fuel cell technology is being pursued in Argentina, at least initially, with a relatively narrow set of applications in mind, the added cost and effort of developing separate benchmarks for different applications appears to be unnecessary. If, as fuel cell technology is introduced in Argentina, the range of applications broadens, then the modified technology matrix can always be expanded to include separate application subcategories for the four main fuel cell technologies.

6.3.2 Benchmark Subcategories

Although it does not appear necessary to sub-classify SOFC technology based on application at this point in time, other benchmark subclassification schemes must be considered. In particular, it is likely that the emission characteristics of the diesel generators that would be used in place of the SOFCs will vary depending on generator capacity. Very small units are likely to be less efficient, and perhaps less well maintained than larger units. For this reason, separate benchmarks for different diesel generator capacity ranges seem warranted. The specific size ranges used to subcategorize the benchmarks should be selected based on an empirical analysis; essentially, the number of subcategories used should be sufficient to capture all significant size-related differences in diesel generator efficiencies.

Note that to use benchmarks classified according to generator capacity, we must assume that the capacity of the counterfactual diesel generators is equal to the capacity of the fuel cells. For example, in our project one 250-kW fuel cell is to be installed. Obviously, in the absence of this project diesel generators of various sizes could be used to meet the 250-kW demand. However, *as a generality* it seems reasonable to assume that diesel generators of roughly the same size would be installed. Certainly, in the case of off-grid applications, this assumption should generally prove valid. For fuel cells serving the grid, the assumption should prove reasonable in some cases although there will, of course, be many exceptions. But once again within the context of the modified technology matrix approach, we are constrained to ignore exceptions and address only the general or typical situation.

Another possible benchmark subcategory is region. It is possible that diesel generator efficiencies will vary by region; for example, climate and altitude may affect the generators' operating performance. However, as in the case of capacity, regional subcategories, if necessary, must be defined based on an empirical analysis of diesel generator operating data. If available data were to show a marked difference in generator efficiencies across different areas of Argentina, then a regional classification scheme could, and should, be developed to capture those differences. However, as in the case for China, the availability of data must be considered in defining regional subcategories.

6.3.3 Benchmark Metric

In the case of the China IGCC project, the average heat rate of recently constructed conventional power plants was selected as the benchmark. In the case of SOFCs in Argentina, we propose that the average emissions rate for new diesel generators (in pounds CO_2 per kilowatt-hour) be used as the benchmark. Because diesel fuel is a relatively homogeneous fuel that can be characterized with a single, constant emissions factor, it is unnecessary to use heat rate as the benchmark. Instead, by multiplying the average heat rate of new diesel generators with the diesel fuel emissions factor, the appropriate benchmark emissions rate can be readily derived. The emissions baseline for any particular SOFC project, in a given year, could then be computed by multiplying the benchmark emissions rate by the amount of electricity (in kWh) generated by the fuel cell(s).

Because any fuel cell project undertaken in Argentina will be a new capacity project (in so far as there are no existing fuel cell projects in Argentina), and because the Argentine government is interested in fuel cells as a means to meet new demand growth in applicable rural areas, we may presume that, in general, an SOFC project will displace new diesel generators as opposed to existing generators. Therefore, the benchmark emissions rates should be developed based on new diesel units. As in China, we propose that units installed within the past five years be used as the basis for the benchmark.

Whenever possible, the benchmark should be based on actual heat rate data for operating diesel generators. It may be possible to obtain such data, at least for larger generators owned and operated by electric utilities. However, for smaller diesel generators owned by the end user, it may well prove difficult, if not impossible, to obtain the required heat rate data from the generator owners. In these cases, it may be necessary to use heat rate estimates provided by manufacturers for specific models, in combination with market share data for the different models. Using this approach, the market share for each model within a particular capacity range would be used to weight the heat rate for the model; the resulting weighted average heat rate would be used to compute the benchmark emissions rate. The market share data must, of course, be limited to generators sold in the preceding five years to ensure that the resulting heat rates reflect the performance of new units.

Finally, it should be noted that some consideration was given to the use of a function rather than a constant as the benchmark. However, use of a function is even less appropriate in the case of Argentine fuel cell projects than it was for Chinese IGCC projects. For one, fuel quality, which is a major determinant of heat rate in the case of coal-fired power plants, is not a factor for diesel generators. Although diesel fuel may exhibit small variations in heat and sulfur content, it is essentially a homogeneous fuel (unlike coal). Furthermore, the chances that the counterfactual will "shadow," or remain synchronized over time with, the project--a key requirement if data for the latter are to be applied to the former--are even poorer in the case of fuel cells than for IGCC units. This follows from the fact that SOFCs will generally utilize natural gas--not diesel fuel. Therefore, as in the case of the Chinese project, the benchmarks will be held constant with respect to all variables except one--time.

122

6.3.4 Temporal Considerations

The heat rates, and emission rates, of diesel generators will vary significantly over time. Heat rates will deteriorate due to normal wear and tear on the generators; the amount of deterioration experienced will depend on the quality of maintenance received by each unit. At the same time, new models will be introduced into the marketplace; these models may improve upon the efficiency of older models.

To ensure that the benchmarks remain a realistic indicator of current conditions, it will be necessary to update them on a regular basis. We propose that the same updating approach developed for the Chinese IGCC project also be applied in Argentina. It will be recalled that, under this approach, a new set of benchmarks, to be applied to new projects *only*, was to be derived every five years. The new benchmarks for SOFC projects would be estimated in the same manner as the old benchmarks: data for diesel generators installed in the preceding five years (or sold in the preceding five years, if manufacturers' data is being utilized) would be used to compute average benchmark emission rates by generator size category and (if applicable) region. These benchmarks would then be applied to all projects initiated before the next five-year update.

At the same time, the benchmarks to be applied to *ongoing* projects would also be updated every five years. In the case of benchmarks derived on the basis of actual operating data for installed units, the updates would be performed in the same manner as for the Chinese project. Each new-project benchmark will be derived on the basis of a selected sample of operating generators; these generators form the benchmark group that will be used as the basis for updating the benchmarks every five years. Using this approach, heat rate data for each selected benchmark group must be collected at least once every five years; these data would be used to compute an average benchmark emissions rate for the group on a five year basis. Once half of the original capacity comprising the group has been retired, any remaining projects benchmarked on the basis of that group will be switched to a new benchmark group (under the assumption that the project counterfactual unit(s) has reached the end of its life).

For benchmarks derived on the basis of manufacturers data, a slightly different approach will be required. A new-project benchmark group will still be identified every five years, but this identification will be based on market share data for the different generator models. For example, if the data show that a particular model, X, comprises 10 percent of the total generators sold in a particular size class in the preceding five years, then that model will be assumed to comprise 10 percent of the benchmark group. This 10 percent value will be held constant over time. Changes in model X's heat rate over time will be based on the manufacturer's estimates. Manufacturers' estimates of the life of model X will be used to determine when to eliminate the model from the benchmark group; once over half of the capacity in the group has been eliminated, the group as a whole will be considered retired.

Because the diesel generators, especially small, off-grid units, are unlikely to be retrofitted, generator efficiencies will probably deteriorate rather than improve over time. Vigilant maintenance can significantly reduce this deterioration, but probably cannot eliminate it. Thus, the emissions rates for each benchmark group are likely to increase as they are updated. The use of benchmarks that increase over time may seem counterproductive given that the ultimate goal of an international climate change agreement would be to reduce emissions. The implications of the alternative must be considered. If the benchmark were to be held constant rather than allowed to increase to reflect real changes in the project counterfactual, the emission reduction credits awarded over the life of fuel cell projects--or for that matter any type of project characterized by increasing counterfactual emission rates--would be reduced. A bias would be introduced into the emission reduction estimation process--the amount of credits awarded to projects would tend to be less than the actual emission reductions achieved by the projects over their lifetime. As a result, developed countries subject to emission reduction goals under an international climate change agreement would in effect "overshoot" the targets--actual achieved emission reductions would presumably exceed the reductions required under an international climate change agreement.

Such an outcome might be considered a positive result--were it not for the fact that it will come at a potentially high cost. By reducing the amount of credits awarded for market-based projects below the level of emission reductions actually achieved, the cost per ton of credits will be effectively increased. This may, in turn, discourage potential project developers from undertaking market-based projects. Consider, in this regard, that the emissions rates of *projects* will often increase over time for the same reason that *counterfactual* emissions rates will increase: normal wear and tear on equipment with the passage of time. Since the project emission rates will be measured on an annual (or other periodic) basis, the increase in these rates will be captured in the emission reduction estimation process. If the corresponding increases in the counterfactual emission rates are not likewise captured, then over time project developers will receive diminishing returns on their investment. This will have the effect of discouraging market-based projects, and forcing developed countries to rely more heavily on other means for meeting (in fact exceeding) their emission reduction targets.

The results will be an increase in real emission reductions achieved under international climate change negotiations, but at a higher cost. This runs counter to the fundamental purposes of the project- or market-based activities. Project- or market-based activities are *not* designed to reduce emissions below the requirements specified under the UNFCCC. Rather, the purpose of such activities is to reduce the *costs* of meeting emission reduction targets to be established under an international climate change agreement, by enabling the transfer of emission reduction activities from developed to developing countries where more cost-effective reduction opportunities are believed to exist. Also, when viewed in a broader perspective, project- or market-based activities may help to put developing countries on a more environmentally sound development path by providing those countries with direct experience utilizing advanced high-efficiency technologies. If the costs of undertaking market-based projects are increased as a result of biases in the emission reduction estimation process, this will not serve the objectives of such activities. Furthermore, the resulting cost increases may have a negative impact on the political will needed to maintain existing commitments, and pursue further commitments, to reducing emissions. In this regard, it is important to recall that one of the key arguments that the United States harbored against the Kyoto Protocol was that it did not require developing countries to make the same commitments required of developed countries. Project- or market-based activities at least partially address this concern by providing some role, albeit voluntary, for developing countries. If the fundamental goal of such activities--cost reduction--is compromised by estimation biases, then project- or market-based activities may come to be viewed as failures within developed countries, and it may become more difficult politically for these countries to commit to further emission reductions. At the same time, the opportunity to introduce new, environmentally-beneficial technologies into developing countries may be lost. The purpose of project- or market-based activities is best served by utilizing an unbiased, accurate approach to emission reduction estimation.

6.4 Summary

In this chapter, we developed the market-based project baseline for a solid oxide fuel cell (SOFC) project in Argentina. This project is an off-grid generation demonstration project, involving the installation of a 250-kW SOFC at a rural village. This project is purely hypothetical, although it is based on the Argentine government's interest in applying fuel cell technology to meet expected growth in the country's rural electricity demand.

As in the case of the China IGCC project, the modified technology matrix approach was selected for the project analysis, and our focus was on developing the matrix itself rather than applying it to the specific project. The additionality of SOFC technology was demonstrated based on a cost comparison between SOFCs and commercial power generation technologies. An assessment of the market penetration of SOFCs, in Argentina and the world as a whole, lent further support to our conclusion that the technology qualifies as additional.

The benchmark for SOFCs was taken as the emission rate of diesel generators installed in Argentina in the last five year. Again, benchmarks for new and existing SOFC projects will be developed on a periodic (five-year) basis, using the "benchmark group" concept introduced in Chapter 5.

In the following chapter, our analysis of the Argentine fuel cell project, and the other two projects, is subjected to a critical assessment.

7. CRITIQUE OF THE PROJECT ANALYSIS

In the preceding three chapters, we approached the project analysis much as project developers (or technology matrix developers) would approach them. We attempted to put forth as strong an argument as possible supporting our conclusions on additionality and the emissions baselines; alternative conclusions and counter arguments were considered only in so far as it was necessary, and possible, to anticipate and refute criticism.

Having thus considered the projects from the subjective standpoint of a developer, we would now like to step back, adopt a more objective viewpoint, and critique the project analysis with a view towards identifying their strengths and weaknesses, and drawing out the lessons that might be learned from this project analysis exercise. There are a number of specific issues that we would like to address:

- Fundamental project analysis difficulties arising from the characteristics of developing economies – More than once during this exercise, we were initially mislead because we applied a Western frame of reference to China and India. There are critical differences between developed and developing countries that hold major implications for market-based project analysis. *The most important of these differences is the existence of chronic supply-demand disequilibriums in the developing world.*

- Weaknesses in the analysis of the Indian Power Plant Efficiency Improvement project – Despite attempts to develop as strong an argument as possible in defense of the additionality of this project, it is quite easy to construct a persuasive counter argument. Furthermore, there is a great deal of uncertainty surrounding the emissions baseline developed for the project.

- Potential for biases in the estimation of project baselines – These potential biases, arising in part from supply-demand disequilibriums, may have the effect of undermining emission reduction targets.

Each of the above issues will be addressed in turn in the following sections. We will conclude with a fourth section (Section 7.4) proposing possible options for addressing the dilemmas identified in Sections 7.1 through 7.3.

7.1 The Implications of Supply-Demand Disequilibriums

Much of the analysis of market mechanisms has been done in the West and has been quite abstract in nature. One of the benefits of considering concrete market-based projects placed in developing countries such as India and China is that it forced us to discern and address issues specific to the developing world that might never arise in the context of a more abstract analysis. The application of general methodologies to specific projects enabled us to see beyond our Western frame of reference and discern certain key characteristics of developing economies that are of great import for the market mechanisms. Some of these characteristics have implications not only for the three

126

projects considered here, but for many of the projects likely to be undertaken as part of the market mechanisms.

The most important of these developing economy characteristics, in terms of its implications for the market mechanisms, is the existence of persistent, chronic supply-demand imbalances throughout much of the developing world. We encountered these imbalances when considering the energy markets in India and China. In both of these countries, demand for electricity and primary fuels (mainly coal) often exceeds supply, as a result of irrational tariff structures, capital availability constraints, and technical problems, such as weaknesses in transport infrastructure. In both countries, demand and supply may come into balance at times; for example, the impact of the Asian financial crisis has reduced energy demand in China leading to a short-term improvement in the supply-demand situation. But over the longer term, the disequilibrium is expected to reappear, because the structural problems underlying the supply-demand imbalance persist. Many other developing countries, besides China and India, face the same problems.

As we have seen in Chapters 4 and 5, the existence of these supply-demand imbalances hold major implications for baseline development. There are two types of imbalances particularly important to the analysis of market-based projects in the power sector: electricity imbalances and fuel imbalances. We will consider electricity first.

7.1.1 Electricity Supply-Demand Imbalances

7.1.1.1 The Indian Power Plant Efficiency Improvement Project. The existence of an imbalance in electricity supply and demand proved to be a major complicating factor in the project analysis, especially in the case of the Indian Power Plant Efficiency Improvement project. As was discussed in Chapter 4, owing to the shortfall in electricity supply in India, it cannot be assumed that a heat rate improvement project will lead exclusively to a decline in fuel consumption. Rather, in most cases, an Indian power plant will use an efficiency improvement to increase the amount of net generation delivered to the grid (thus indicating a need to utilize a modified unit of production baseline). There are technical constraints ultimately limiting gross generation; however, to the extent that an efficiency improvement project relaxes these constraints, or returns them to original design levels, gross and net generation can be increased. In contrast to the situation in India, a heat rate improvement project will almost always lead exclusively to a reduction in fuel consumption in the United States, because generation is determined not by supply but by demand (which will remain unaffected by the project).

The use of heat rate improvements to increase generation hold two main implications for the project-specific approach to baseline estimation. First, it is much harder to consider the development of a standardized set of baseline estimation algorithms, even for application to the relatively narrow range of projects that could be considered heat rate improvements. For one, electricity supply-demand imbalances characterize many, but not all, developing countries. Furthermore, even in countries plagued by such imbalances, the extent to which a particular heat rate improvement project affects generation as opposed to fuel consumption will vary greatly depending on the characteristics of the power plant as well as the project specifics. A power plant that is operating at design levels will not be able to increase its generation irrespective of an efficiency improvement. On the other hand, a project that focuses exclusively on certain plant auxiliary equipment (such as pumps) may have no affect on fuel consumption while enabling all of the resulting electricity savings to be delivered to the grid in the form of increased net generation. Between these two extremes lies a full range of potential project impacts on fuel consumption and net generation; assessing the change in each of these two variables will require the development of power plant measurement and monitoring protocols that may need to be tailored to each individual plant.

Finally, the emission sources displaced by the increased generation may differ from country to country, or, for that matter, from one region of a country to another. In India, commercial establishments utilize diesel generators during the rolling blackouts caused by the electricity supply shortfall. In other countries, supply shortfalls may take the form of brownouts during peak load periods; or, if they are manifested as rolling blackouts, they may affect other emission sources, such as kerosene lamps or heaters, coal furnaces, wood-burning stoves, etc. Furthermore, the specific kinds of data that may be available on these various end-use emission sources may vary greatly from one country to another, making it difficult to specify a single procedure or set of algorithms applicable to all situations.

As was suggested in Chapter 3, the development of standardized procedures for the project-specific approach may run counter to the goals and purpose of this approach. In the project-specific approach, the basic goal is to account for the key project-specific factors affecting emission reductions. To the extent that these factors share similarities within groups of like projects, it may be possible to develop standard procedures/algorithms. However, our analysis of the Indian Power Plant Efficiency Improvement project illustrates the dangers inherent in any attempt to standardize the approach. There may simply be too many critical factors, varying in too many ways across different projects, to enable the development of standard procedures. The existence of electricity supply-demand disequilibriums, in particular, has the potential to manifest itself in myriad ways for different projects and countries. If standard procedures were to be developed and applied, there is a strong possibility that the resulting reduction estimates will fail to account for critical differences across projects. As the name implies, the project-specific approach may ultimately require the development of a unique set of algorithms for each different project.

The second major implication of generation increases following upon heat rate improvements has to do with the level of uncertainty surrounding the baseline estimate. In the United States and other developed countries, the impact of a heat rate improvement project on emissions is limited primarily to the power plant directly affected by the project. However, owing to supply-demand imbalances, this will not be the case in India and other developing countries. In these countries, the direct impact of a heat rate improvement at a single power plant may extend to hundreds of thousands of small end-use emission sources, ranging from large coal-fired boilers at industrial establishments to kerosene lamps used by individual households. As a result, the development of an accurate emissions baseline is *much* more difficult for developing countries than for developed countries. Merely identifying the types of emission sources that may potentially be affected by a heat rate improvement project would appear to be a daunting exercise. Obtaining reliable data on these sources may, in many cases, prove impossible. At best the project analyst must rely on averages and overall estimates to characterize the emission rates of groups of similar emission sources; there is no hope of developing an inventory identifying and characterizing each affected source. Furthermore, the available data, at least for the smaller emission sources used in the residential and commercial sectors, is likely to take the form of manufacturers specifications and estimates. Operational efficiencies of devices, such as diesel generators, may differ markedly from manufacturers' estimates, especially as the devices age. In fact, manufacturers have an incentive to exaggerate the efficiencies of their products; thus, there is a strong possibility that the use of manufacturers data will bias the baseline emission estimates (and hence, the emission reduction estimates) towards the low side. Conducting surveys of a representative sample of the affected emission sources may be a means of obtaining better data, but such surveys would themselves be fraught with difficulties. Particularly for residential and small commercial electricity customers, survey response rates could prove very low, and the reported data could prove highly unreliable. Ultimately, the only way to ensure the collection of an adequate amount of reliable data might be to conduct field testing and monitoring of a representative sample of emission sources. This would be a very expensive undertaking.

Another possible approach to the problem might be to ignore the potential for generation increases resulting from heat rate improvement projects, and simply assume that such projects result exclusively in a decrease in fuel consumption. In other words, we could adopt, as a convention, the emissions baseline development approach that would be used in developed countries, and apply it to market-based efficiency improvement projects in developing countries. The adoption of such a convention would significantly simplify the baseline estimation process, by limiting the data requirements to the power plant directly affected by the project. Presumably it should be possible to obtain better, more reliable data on a single affected power plant than on thousands of small end-use emission sources. Of course, the drawback to this approach is that it relies on a convention that may well be false. Furthermore, because the emissions characteristics of coal-fired power plants are likely to differ, systematically, from those of diesel generators, kerosene lamps, and other end-use emission sources, the adoption of a convention ignoring the latter sources may introduce significant biases into the resulting emission reduction estimates.

With the possible exception of a survey of end-use sources involving field monitoring and testing, the other alternatives are also likely to produce unreliable, biased estimates. This being the case, it is not clear that these more costly alternatives are preferable to simply assuming that generation remains unaffected by heat rate improvements. There is no question that uncertainties in the emission reduction estimates, and especially biases in the estimates, represent a significant potential problem. However, attempts to eliminate all biases may prove very expensive, and may fail despite their expense. *Even if they succeed, costly solutions to the problem of biased estimates bring with them their own problems: high project development costs will discourage project developers from undertaking market-based projects, and run counter to the basic purpose of the market mechanisms, which is to reduce the costs associated with emission reduction estimation activities.* Rather than attempting to eliminate biases through costly data gathering techniques, there may be an alternative. We will return to the issue of market mechanism estimation biases, and possible options for accommodating these biases within a future emission reduction agreement, later in this chapter.

7.1.1.2 The Chinese IGCC Project. Electricity supply-demand imbalances also affected our analysis of IGCC technology in China, although the impact was not as great in China as in India. The main

129

reason for this difference was that in the case of China, we were considering projects involving the construction of *new* capacity to meet new demand. In the United States and other developed countries, it is fairly safe to assume that new demand will be met by new capacity; even if a particular capacity expansion project is abandoned for one reason or another, it will be replaced by another project assuming the demand remains in place. However, because of the electricity supply shortfall in some of the fastest growing regions of China, we were not able to assume that new demand would be met by new conventional capacity in the absence of the project. Rather, capital availability is a key supply constraint in China and other developing countries, and unless we could show that abandonment of the IGCC project would free up more capital to pursue another capacity expansion project(s), we would have concluded that the demand to have been met by the project would have remained unmet.

In the case of Chinese IGCC technology, we were able to argue that, *in general*, the capital freed up in the event of project abandonment would *probably* be used to support the construction of conventional coal-fired capacity. Thus, our treatment of the IGCC project did not differ materially from how we would have treated it were it located in a country without electricity supply-demand imbalances. However, the existence of these imbalances in China and other developing countries adds one more level of uncertainty to an already highly uncertain analysis. Perhaps for some projects, the project financing would be diverted to other sectors of the economy, or other countries, if the projects were to be foregone. Particularly in the context of the modified technology matrix, our analysis is limited to a consideration of the "typical" project; clearly, there may be many actual projects that are atypical and that do not fit the general conclusions embodied in the matrix. The true counterfactual for these projects may differ markedly, and systemically, from our assumed counterfactual.

7.1.2 Fuel Supply Demand Imbalances

In addition to electricity supply shortfalls India faces significant, persistent coal supply-demand imbalances. Power plants often must cut back on generation owing to a lack of fuel. Other developing countries experience similar shortages in coal and other fuels. These shortages arise as a result of a variety of factors, including for example, lack of capital to expand domestic mine capacity, rail transportation bottlenecks, and port capacity constraints.

The potential implications of these fuel shortages for market-based projects are even greater than the implications of electricity supply shortages. We touched on these implications in Chapter 4, in our discussion of the secondary effects of the Indian Power Plant Efficiency Improvement project. To reiterate the basic concern, if the fuel demands of power plants are going unmet, then market-based projects designed to reduce power plant fuel demand may have *no impact* on fuel consumption or emissions. Rather, such projects (including, for example, all efficiency improvement projects) may simply reduce the size of the supply-demand gap. In countries facing fuel supply shortages, fuel consumption is determined not by demand but by supply constraints; projects that reduce demand without addressing the supply constraints will have no impact on total emissions. Such projects will reduce emissions per unit of output; however, the awarding of credits must be based on an absolute, not a relative, measure of reductions.

The existence of chronic fuel supply shortages in developing countries holds major implications for emission baseline estimation. Given these shortages, the emission effects of projects, such as the Indian Power Plant Efficiency Improvement project, may affect only the emissions of small end-use generators; power plant emissions may remain unchanged despite the reduction in coal demand. The Chinese IGCC project might even be worse, in that it might have no effect on emissions because the coal saved as a result of this project might simply be consumed elsewhere.

As noted in Chapter 4, the situation is not as simple as it appears. The magnitude of the supply shortfall varies significantly over time and from one power plant to another; a detailed analysis of the supply shortage would be required to estimate its effect on the proposed market-based projects. Such an analysis would be a difficult, expensive undertaking and fraught with uncertainty. The alternative would be to simply ignore the potential impact of supply shortages on emission reduction estimates. This alternative would almost certainly result in erroneous, and biased, emission reduction estimates. Specifically, emission reductions would tend to be overestimated; the amount of emission reduction credits awarded would exceed the reductions achieved; and overall global reductions would fall short of emission reduction goals. Once again, we are faced with a choice between requiring the use of a difficult estimation procedure that will drive up the cost of market-based projects and discourage developers from undertaking them, or accepting biased reduction estimates that may undermine overall reduction goals. Later in this chapter we will return to the issue of biased estimates and consider two possible options for dealing with them: (1) eliminating them through the application of very stringent project analysis techniques, and (2) accommodating them through the use of adjustable emission reduction targets.

7.2 Weaknesses in the Analysis of the Indian Power Plant Efficiency Improvement Project

7.2.1 Additionality

The Indian Power Plant Efficiency improvement project presented particular difficulties from an analytical standpoint. Some of these difficulties stemmed from the fact that the project involves relatively minor modifications to power plants that utilize conventional technology. Whereas the additionality of the technologies utilized by the Chinese and Argentine projects is quite clear cut and relatively easy to establish, this was not the case for the Indian project. Although we *believe* that the latter project is additional, *demonstrating* its additionality is more difficult. Because the project is designed to produce significant efficiency gains at low cost, and to be widely replicable at coal-fired power plants throughout India, we were unable to utilize the economic feasibility approach to establish the project's additionality. Instead, we were forced to rely upon the non-economic barriers approach. Although we approached the project analysis from the subjective standpoint of project developers, and attempted to craft as strong an argument as possible in support of the project's additionality, both the financial barrier argument and the knowledge barrier argument are weak in certain key respects.

7.2.1.1 The Financial Barrier Argument. The financial argument hinges on our assessment of the weak financial position of the Indian State Electricity Boards (SEBs). Essentially, we argue that because the SEBs are operating at a significant loss, they would not be able to raise the capital required to pursue the efficiency improvement project on their own. Because they require the financial aid of the U.S. partners, we conclude that the project could not have occurred without the market mechanism incentive provided for U.S. participation in the project.

Although this is a reasonable, and reasonably strong argument, a persuasive counter argument can easily be constructed. If, as we conclude, the SEBs weak financial situation would have precluded them from undertaking the project on their own, then their financial weakness would, by the same logic, prevent them from undertaking *any* project. The basic weakness in our financial barrier argument is that it is not specific to the project at hand. We essentially argue that the SEBs could not undertake the project on their own because their financial position precludes them from raising the needed capital; by this same argument, we would have to conclude that the SEBs are unable to undertake *any* capital improvement project. Yet clearly the SEBs can, and do, undertake some projects; why then could they not undertake the efficiency improvement project?

To strengthen our argument, a key component is required: it must be shown that efficiency improvement projects are a relatively low priority within the SEBs. If it were possible to argue that this is the case, we could then argue that the SEBs are only able to raise the capital needed to fund top-priority projects; the conclusion could then be drawn that the available capital is insufficient to fund a low-priority project such as the Indian Power Plant Efficiency Improvement project.

Unfortunately, although we attempted to find support for the position that efficiency improvement is a low priority within the SEBs, the available evidence in fact suggests the opposite. In the government's current five-year plan, efficiency improvements at existing power plants are identified as a top priority. Because efficiency improvement projects might be used to increase generation, they may be viewed as a low-cost alternative to the construction of new capacity. Given India's rapidly-growing electricity demand and capital availability constraints, efficiency improvements at existing power plants may well be a higher priority than new capacity projects. The SEBs are quite interested in pursuing efficiency improvements, as evidenced by the fact that some of them have expressed an interest in participating in the GEP project--the ongoing project that forms the basis for our hypothetical project.

The weaknesses in our argument, as developed for our particular project, should not be misconstrued as weaknesses in the financial barrier approach itself. As the above discussion suggests, this same approach could be used to construct a stronger argument in other circumstances –particularly for projects that can be shown to be a relatively low-investment priority. The financial barrier approach, and for that matter the barrier approach in general, can probably never rise to the same level of persuasiveness as the economic feasibility approach. However, while the former approaches will probably always fall short of proving or demonstrating additionality beyond all reasonable doubt, they may suffice for the construction of a sufficiently strong argument in at least some cases. Unfortunately, the particular circumstances surrounding the Indian Power Plant Efficiency Improvement project do not lend themselves to the development of a strong financial barrier argument.

7.2.1.2 The Knowledge Barrier Argument. The weaknesses in the knowledge barrier argument may be more a function of the approach itself than of the particular circumstances surrounding the project. The basic tenet underlying this approach is valid: an organization cannot undertake a project that they do not know how to undertake. However, the practical application of this approach requires the demonstration of a negative, and a particularly difficult negative at that. Is it really possible to persuasively argue that a large organization, such as India's National Thermal Power Corporation (NTPC), lacks the technical knowledge necessary to undertake a particular project? Our argument relied quite heavily on general observations and impressions concerning the state of knowledge within the NTPC; these observations and impressions are based on limited interactions with some NTPC personnel. Given their anecdotal nature, the observations and impressions garnered by NETL and contractor staff over the course of the GEP project clearly fall far short of a demonstration of an organization-wide lack of knowledge. Perhaps the most persuasive evidence provided in support of the knowledge barrier was the data on the number of hours of training received by NTPC staff as part of the GEP project. At least this evidence rose above the level of subjective impression. Unfortunately, this information, while available to us in the context of our hypothetical project, might not be available to developers of an actual project. Clearly, developers submitting a future market-based project for approval will have no advance data on the hours of training ultimately provided to project personnel.

In short, the knowledge barrier argument is quite weak because it relies on generalities, impressions, and a limited amount of data that might not be available to project developers in actual circumstances. It might be possible to strengthen the knowledge barrier argument, although gathering the hard evidence needed to bolster the argument would be a costly, if not impossible, undertaking. One possible approach would be to conduct organization-wide interviews of SEB personnel, as a means of ascertaining their knowledge of the technologies and procedures comprising the project. However, interviewees would have a significant incentive to reveal less than they know about the project, in so far as project approval would depend on the demonstration of a lack of knowledge. Furthermore, it may not be possible to ascertain overall corporate knowledge on the basis of the knowledge of the individuals comprising the corporation. For example, while a number of individuals may have knowledge of only certain specific components of the project, these same individuals might *as a group* possess all of the knowledge required for project implementation.

7.2.1.3 Some Conclusions Concerning the Additionality of the Indian Power Plant Efficiency Improvement Project. None of the above is meant to suggest that the hypothetical Indian Power Plant Efficiency Improvement project is not additional; rather the problem is that it is difficult to demonstrate the project's additionality. We might think of market-based projects as falling into three distinct groups. First, there are some projects, including, for example, the China IGCC project and the Argentine fuel cell project, that are clearly additional; demonstrating the additionality of these projects is a relatively easy and straightforward process. Second, there is a group of projects that are clearly not additional, including, for example, heat rate improvements resulting from standard maintenance activities or new capacity projects utilizing purely conventional technologies. And finally, there is a third group of projects – some additional, and some non-additional – that are difficult to assess from the standpoint of additionality. The Indian Power Plant Efficiency project is one example of this "borderline" group of projects.

Should the Indian Power Plant Efficiency Improvement project be granted approval under the market mechanism? Although there are serious flaws in the arguments presented in support of the project's additionality, the arguments should not be completely dismissed. They do not rise to the level of an unquestionable demonstration, but they nonetheless provide significant evidence supporting the project's additionality. Yet again, we are faced with a difficult tradeoff. On the one hand, if we relax the requirements for demonstrating additionality to enable the Indian project and other additional borderline projects to gain approval, then it is likely that many non-additional borderline projects will also gain approval. Furthermore, because non-additional projects will tend to be more attractive than additional projects from an economic standpoint, investment dollars will tend to flow into the non-additional projects at the expense of the additional projects. As a result, emission reduction credits will be awarded to projects that fail to reduce emissions, and emission reduction goals will be undermined.

On the other hand, if the requirements for demonstrating additionality are tightened to the point where the borderline projects are effectively excluded, then market-based project developers will be left with a much more limited set of relatively unattractive investment opportunities. Many of the most cost-effective emission reduction opportunities will reside within the group of borderline projects, while, in many cases, the projects that are clearly additional *are* clearly additional precisely because they utilize high-cost, high-risk technologies. Limiting investment opportunities to this group of projects will discourage participation in market-based projects, and will drive up the costs of meeting emission reduction goals. We will return to the dilemma presented by these two problematic alternatives later in the chapter.

7.2.2 The Emissions Baseline

The additionality arguments are not the only components of the Indian project analysis that present difficulties; major problems are also apparent in the analysis of the project baseline. Some of these problems have already been discussed. The presence of electricity supply-demand imbalances in India significantly complicated our analysis, and may necessitate the use of difficult-to-obtain, highly-suspect data on thousands of small end-use diesel generators. Similarly, Indian coal shortages raise the possibility that the coal demand reductions caused by the project may not necessarily lead to reductions in coal consumption. These are far from the only uncertainties surrounding the emission baseline. In addition to the uncertainties surrounding the data for small end-use generators, there are major questions surrounding the power plant data that are also required to solve the baseline algorithms. Particularly in developing countries, such as India, power plant fuel consumption and fuel quality data may not be accurate. For example, the data available from the SEBs may prove to be significantly less reliable than NTPC data, simply because the SEBs lack the capital necessary to maintain and upgrade their measurement equipment, fuel sampling programs, and laboratories.

Perhaps of greater concern than these data uncertainties are the uncertainties surrounding our identification of the qualitative baseline. It is extremely difficult to assess what might have happened in the absence of the project. We *believe* that the particular counterfactual we developed for the project was the most likely of the various possible alternatives, but given the nature of the problem there is no way to test or prove this belief. Furthermore, in developing the hypothetical project, we made a key modification to the GEP project, to simplify the baseline development exercise. Specifically, we assumed that the financial contribution of the SEBs would be small relative to the contribution of the U.S. partners. By adopting this simplifying assumption, we were able to conclude that, if the project were to be foregone, the capital thus freed up within the SEBs would be insufficient to support alternative projects. In fact, the NTPC is contributing the majority of the funding--$10 million--to the GEP project. Thus, in the case of the actual project upon which our hypothetical project is based, it is quite possible that capital *would* have been freed up to pursue alternative projects had the efficiency improvements not been undertake. Determining the baseline for the actual project thus would have been a far more complex exercise than the one documented in Chapter 4. How would an analyst identify the potential alternative projects the NTPC might have pursued absent the GEP project? The possibilities are virtually endless, ranging from power plant availability improvement projects that, given India's electricity shortfall, might cause emissions to increase, to training programs or computer upgrades that would, at best, have an indirect emissions effect that would be extremely difficult to quantify. Assuming an analyst could somehow identify the myriad possibilities, how would he or she determine the most likely project alternative(s) from among these possibilities? And then, how would the analyst estimate the impact of the alternative project(s) on emissions? How much would it cost to undertake the rigorous analysis required to address these issues? In the end, would the final baseline estimate justify the cost, or would it be so uncertain and unreliable that the added cost involved in such a rigorous project-specific analysis would have been wasted?

7.3 Qualitative Error Assessment

In the preceding section, we focused on the Indian Power Plant Efficiency Improvement project, but our analysis of the other two projects are also no doubt subject to a high degree of uncertainty. Our conclusions concerning the additionality of IGCC and SOFC technologies are probably much more reliable, or at least persuasive, than our additionality analysis of the Indian project. Additionality is simply a much more clear-cut issue in the former cases than in the latter case: the Indian project falls in the "borderline" group of projects with respect to additionality, while IGCC and SOFC technology are clearly non-commercial, and hence, additional at this point in time.

However, the estimation of the benchmark for these two technologies may well be as prone to uncertainty as the estimation of the baseline for the Indian project. The uncertainties are simply revealed more clearly in our analysis of the latter project, because of the highly-detailed nature of the project-specific approach. This bottom-up approach to baseline estimation tends to reveal potential error sources that may remain hidden to analysts using the less-detailed, top-down modified technology matrix approach. But let us be clear: many of the same types of potential errors discussed above in the context of the Indian project could also affect the Chinese and Argentine projects. For example, coal shortages in China raise the same possibility that exists in India: that projects designed to reduce coal demand may ultimately have no effect on coal consumption. The problems associated with obtaining reliable data on small end-use generators are the same for the Argentine fuel cell project as for the Indian project. The fundamental uncertainties surrounding the establishment of the qualitative counterfactual arise for all three projects. In Argentina, it is true that the government is currently pursuing fuel cell technology as an alternative to diesel generators, but is this how fuel cells will ultimately be used in Argentina? And in China, would the capital required to build a conventional coal-fired plant be freed up in the absence of the IGCC project? Or would the demand simply go unmet?

On top of these uncertainties, the modified technology matrix approach adds a new source of potential errors to the China and Argentina projects that did not arise in the context of the Indian project. The modified technology matrix approach involves the application of a single benchmark, developed on the basis of the "typical" project utilizing a particular technology, to all projects utilizing that technology. Clearly, in so far as actual individual projects differ from the assumed typical project, errors in the baseline estimates may arise.

The single most important "lesson learned" from our project analysis exercise, or at least the lesson that most impressed itself upon us, is the sheer difficulty of developing a realistic, credible, and reliable baseline or benchmark estimate. The potential sources of errors are more numerous, and the potential magnitude of these errors much larger, than originally imagined. Potential errors abound at every step of the analysis. The establishment of additionality using the non-economic barriers approach is fraught with difficulty. The identification of the most likely project alternative is a very difficult and highly uncertain undertaking. Economic circumstances common in the developing world, such as electricity supply-demand imbalances, significantly complicate the process and necessitate the consideration of emission sources far removed from the power plant(s) directly affected by the project. The data required to characterize these sources may not exist, and are likely to be highly unreliable if they do exist. The secondary effects of fuel demand reduction projects, in China, India, and other countries facing chronic fuel supply shortages, may be so large that they effectively cancel out the projects' primary effects. Yet identifying and measuring these secondary effects will require difficult, costly, and error-prone analysis of fuel supply constraints, including assessments of transportation bottlenecks, coal mine capacity and capacity constraints, as well as regional, local, and temporal analysis of the magnitude of the supply-demand gap. Defining a "typical" project and project counterfactual for the purpose of developing the modified technology matrix is a difficult but necessary process, yet one that ensures errors when the matrix is applied to atypical projects.

7.3.1 Biases

7.3.1.1 Biases Arising from Baseline Estimation. If there was reason to believe that the estimation errors will be randomly distributed across market-based projects, the number and magnitude of the individual errors would be less a cause for concern. Given a random or normal error distribution, we could expect the errors to cancel each other out at the global level. Unfortunately, at least some of the identified potential errors are likely to prove systemic. For example, the use of manufacturer's data for small end-use emission sources may prove necessary, yet such data is likely to prove biased. Other types of data may also prove to be highly biased for political or other reasons; for example, it is widely believed that coal consumption in China may be underestimated by as much as 200 million tons; under-reporting of consumption to escape tax implications is widespread throughout the country. The impacts of secondary effects – and especially fuel shortage effects – on projects are also

136

likely to prove systemic; hence, procedures that fail to adequately address these effects can be expected to yield biased estimates.

7.3.1.2 Biases Arising from Additionality Classification Errors. Thus far, we have considered only the possibility of biases arising from the baseline estimation process, but analysis of a project's additionality is a much greater potential source of biases. In fact, purely *random* errors in the classification of projects according to their additionality status can be expected to lead directly to *biases* in emission reduction estimates. For example, if a particular additionality analysis technique leads to the mis-classification of a certain number of non-additional projects as additional, and an equal number of additional projects, with the same emission reduction estimates, as non-additional, the analysis technique will lead to an overestimation of total emission reductions. This follows from the asymmetrical character of the results following upon additionality mis-classifications. *If a non-additional project is approved as additional, it will be undertaken, and it will be awarded emission reduction credits. However, if an additional project is mis-classified as non-additional, it will not be undertaken, because by definition an additional project is a project that will not be implemented absent the awarding of emission reduction credits. Hence, when an additional project is mis-classified as non-additional, the costs of meeting emission reduction goals increase due to the loss of a (presumably) low-cost emission reduction opportunity. However, the estimation of total emission reductions remains unaffected by the error. However, when a non-additional project is mis-classified as additional, emission reductions are overestimated and global reduction efforts consequently fall short of reduction targets.* This asymmetry, which arises from the very nature and definition of additionality, ensures that even randomly distributed classification errors will lead to biased emission reduction estimates.

Furthermore, for all projects that are ultimately implemented, additionality classification errors *always* lead to emission reduction estimation errors equivalent to 100 percent of the estimated project reductions. In other words, additionality mis-classification errors lead to estimation errors that are highly systemic and, at least at the project level, very large in magnitude. Added to this is the potential for even further biases, arising from the likelihood that project developers will preferentially fund non-additional projects mis-classified as additional over truly additional projects, because the former will tend to be more economically viable. For these reasons, additionality classification errors are of fundamentally greater concern than all but the most egregious baseline estimation errors.

7.4 Towards a Solution

This report is by no means the first to emphasize the difficulty of project baseline estimation. Nor is it the first to illustrate the difficulties using concrete examples. Recent analysis of Joint Implementation projects have likewise provided concrete examples of the difficulty of project baseline establishment and emission reduction estimation. However, these analysis have focused primarily on the high costs associated with baseline development, while failing to address an equally important concomitant of the difficulty of baseline estimation: large, systemic potential errors. Thus, the benchmarking approach has emerged as the main proposed solution to the difficulties associated with baseline development. While this approach will unquestionably reduce the costs associated with baseline estimation, it may also significantly increase the potential for additionality classification errors. By allowing such clearly non-additional projects as those involving existing or new hydroelectric and nuclear facilities to qualify for credit, the benchmarking approach has the potential for undermining emission reduction goals.

Ideally, a balanced solution to the difficulties posed by project baseline estimation should address *both* the high costs of the estimation process *and* the high potential for estimation errors and biases. In the following sections, we will consider two possible approaches to error and cost reduction: (1) reducing biases through the development and application of rigorous baseline development procedures and (2) accommodating biases within the framework of future international agreements. We will begin by considering error reduction techniques for additionality classification errors.

7.4.1 Reducing Additionality Classification Errors

In general, reducing biases should be preferable to accommodating biases, as long as the error reduction techniques are not overly costly. This is particularly true in the case of additionality classification errors, because, as we have seen, these errors will lead to large, systematic biases in emission reduction estimates. Hence, cost-effective approaches for reducing these errors are well worth considering.

The modified technology matrix may be one such approach. It is essentially a modified version of the benchmarking approach, but with the addition of an effective additionality screen based on project technology. Because it focuses on advanced, non-commercial technologies, the matrix's additionality screen can be developed fairly easily, and at limited cost. In fact, qualifying an advanced technology for inclusion in the matrix is considerably easier than developing the benchmark for the technology. The relative ease of qualifying advanced technologies is illustrated by the Chinese IGCC and Argentine Fuel Cell projects. Furthermore, once a technology has been qualified, assessing the additionality of any particular project using that technology becomes a trivial exercise. The project developers need only demonstrate that their project utilizes the qualified technology to receive automatic project approval. One drawback of the modified technology matrix is that, unlike the benchmarking approach, it is limited in scope to only those projects employing qualified technologies. However, for these projects it should dramatically reduce the costs of project analysis, while at the same time screening out other non-qualifying projects that might be misclassified as additional under the benchmark approach.

Options, such as the modified technology matrix, that can significantly reduce the potential for biases in a cost-effective manner should be given careful consideration. The modified technology matrix meets the twin goals of cost and systematic error reduction and may hence be preferable to alternatives, such as the benchmark approach (which meets only the cost reduction goal).

However, in order for it to be an effective additionality screen, the developers of the technology matrix must use stringent technology qualification criteria. Furthermore, because the project-specific approach must be used in combination with the modified technology matrix, to cover projects that utilize non-qualifying technologies, the additionality tests applied under the project-specific approach should also be rigorous for the sake of logical consistency. Less-rigorous additionality criteria could, of course, be used to reduce project development costs. However, the use of the modified technology matrix in conjunction with loose additionality testing would offer little if any benefit over the benchmark approach. Therefore, if the benefits of reduced costs are judged to outweigh those of reliable emission reduction estimates, the benchmark approach probably represents a better option than the use of the combined technology matrix/project-specific approach.

However, higher reliability versus lower costs may perhaps be a false choice. Although it may seem that the use of rigorous additionality tests will drive up project development costs, this is not necessarily the case. In fact, the costs associated with stringent additionality criteria may not differ much from the costs of utilizing loose criteria.

This is not to say that the costs of using stringent criteria will not be high. Clearly, in the case of the project-specific approach, the requirement that developers provide a rigorous demonstration of additionality may significantly increase the costs of project analysis. Also, although project additionality analysis costs will remain low in the case of the modified technology matrix irrespective of the stringency of the criteria used to qualify the technologies, other costs will be incurred. In fact, the costs associated with "lost opportunities" – i.e., additional projects that will be classified as non-additional because they cannot meet the stringent additionality criteria – are of greater importance than the project analysis costs. These lost opportunity costs will be quite high. Clearly, additional projects utilizing advanced technologies will readily qualify even given rigorous additionality tests, but many borderline additional projects will not be able to meet the demand for a clear demonstration of additionality. Yet, it is precisely these borderline additional projects that will usually provide the most cost-effective emission reduction opportunities. These projects will tend to be more economically viable than those utilizing advanced technologies; in fact it is for this reason that their additionality status will be difficult to demonstrate. Some borderline additional projects should still be able to qualify as additional even given a rigorous project-specific additionality test, but many such projects will not qualify. Rigorous additionality screening will unquestionably drive up the costs of market-based project development, and it will likely discourage participation in the market mechanisms and reduce its effectiveness as a means of lowering the costs of meeting emission reduction goals.

But while lost opportunity costs will be high given rigorous additionality screening, they will also be high given loose screening. If the additionality criteria are relaxed, then it is true that more borderline additional projects will qualify for crediting. However, more borderline *non-additional* projects will also qualify. *Also, because non-additional projects will tend to be better investment risks than additional projects, project developers will tend to fund borderline non-additional projects at the expense of additional projects.* Non-additional projects are, by definition, "business-as-usual" projects that do not require the added incentive of emission reduction credits in order to be undertaken. Additional projects, by definition, require these added incentives. Given a situation in which both types of projects will receive the credits, investors will tend to prefer the lower-risk, business-as-usual non-additional projects. Hence, although they may qualify more readily for credits under a loose set of additionality criteria, borderline additional projects may be foregone in favor of non-additional projects given such criteria. Again, some such projects will probably be undertaken, but many may be foregone. *In short, it appears that borderline additional projects will to a large extent be lost opportunities regardless of the stringency of the additionality rules.* Given stringent rules, they will fail to qualify for crediting; given more relaxed rules, they will be foregone in favor of non-additional projects.

Hence, the lost opportunity costs associated with rigorous additionality screening should not differ much from those associated with loose screening. It is true that project analysis costs will be lower in the latter case, but it is quite possible that non-additional rather than additional projects will largely benefit from the lower costs. At the same time, the misclassification of large numbers of borderline non-additional projects may bias emission reduction estimates and lead to a shortfall in efforts to meet emission reduction goals. To the extent that market-based projects exert an influence on overall technology decisions made in the host countries, the qualification of large numbers of non-additional projects may encourage developing countries to follow a "business-as-usual" development path. Ultimately, the misclassification of large numbers of non-additional projects as additional may undermine the credibility of the market mechanisms.

7.4.2 Accommodating Biases in Emission Baseline Estimates

In the case of additionality, it may thus be possible to reduce classification errors in a cost-effective manner, by utilizing the modified technology matrix approach and applying stringent additionality criteria. However, in the case of at least some of the types of systematic errors expected to arise during baseline (or benchmark) estimation, it appears that there may be no cost-effective error reduction techniques available. For example, the only reliable alternative to utilizing potentially biased manufacturer's data on small end-use generators may be to perform field surveys of actual operating units. Such surveys would be an expensive undertaking. Similarly, detailed regional and temporal analysis of fuel supply-demand imbalances would be required to accurately capture the impact of such imbalances on individual demand reduction projects.

Diligent, costly error reduction techniques may also not succeed. In fact, although such techniques may reduce systematic errors, it seems virtually impossible that all such errors could be eliminated. For one, it is very difficult to identify all potential sources of systematic errors, and impossible to know whether and when all such sources have been identified. We have discovered a few potential systematic error sources in the three example projects; we suspect that there may be more. Furthermore, the steps that must be taken to reduce errors may themselves become sources of further errors. For example, analysis of fuel supply-demand gaps and other secondary project effects will add further levels of complexity onto an already complex problem.

In short, attempting to deal with systematic errors by eliminating them will likely prove very costly, and may at best only partially succeed. Certainly, in some instances, it may be possible to reduce errors in a cost-effective manner; in all such cases the required error reduction techniques should be utilized. But in other cases, where attempts to reduce errors involve highly costly, complex analysis that may themselves become sources of further error, might there not be an alternative to error reduction?

7.4.2.1 Placing International Agreements in Perspective. One possible alternative would be to find some way of accommodating systematic errors within a future emission reduction agreement framework. *Here, it is important to place an international emission reduction agreement in perspective. An international agreement will not, and cannot, solve the problem of global climate change in and of itself.* Even assuming that all signatories to an agreement comply in full, emissions from non-signatory developing nations will continue to rise rapidly. Atmospheric concentrations of carbon dioxide and other greenhouse gases will likewise continue to rise. An international agreement may slow this process down to some extent, but it will not reverse it. Mitigating global climate change will require a long-term effort by both developed and developing countries; the Kyoto Protocol for example represents but a first step in this effort.

However, unless it is perceived as an abject failure, it is likely that market mechanisms will in one form or another continue to be a part of future agreements, unless all developing countries agree to mandatory emission restrictions. Perhaps full developing-country participation will be achieved in the next major international agreement. If this occurs, then trades between developed and developing countries will be in the form of emission allowances rather than emission reduction credits. Market mechanisms will have no role to play in this situation, and the difficulties and costs inherent in project-level analysis will be rendered moot. However, full participation in a future mandatory emission reduction scheme may be a long time in coming. In the meantime, market mechanisms will likely remain the instrument by which non-Annex I countries participate in emission reduction efforts.

This being the case, might there not be some way of accommodating market-based emission reduction estimation errors within the framework of future agreements? One simple, albeit crude, way to accommodate errors would be to agree upon mandatory emission reduction targets that include some added "cushion" to account for systematic errors expected to arise from market mechanisms. For example, if the ultimate goal is to reduce global emissions by x tons, then the actual agreement could call for a total reduction of x+e tons, where e represents some estimate of the total error expected to arise from market-based projects (and, for that matter, from other sources such as national emission inventories). It should be possible to obtain at least a rough sense of an appropriate value for e based on our experience with market-based mechanisms. For example, we might estimate the expected effect of Kyoto, under the assumption that no emission reduction estimation errors will occur, on global emissions or atmospheric greenhouse gas concentrations. Then, we could compare these expectations against actual emission or atmospheric concentration levels; the differences between the expected and actual levels would provide at least a rough indication of the combined impact of market mechanism emission reduction errors, other emission estimation errors, and compliance failures.

The Need for Accommodating Market Mechanism-Related Errors in Future Agreements

The need for some means of accommodating potential errors is made manifest by the uncertainties inherent in any attempt to identify a "what if" alternative. For example, as we have seen in the case of the Indian project, the counterfactual might have been any of a myriad number of potential alternative projects. How would an analyst identify the possible alternatives? The possibilities are virtually endless, ranging from new capacity projects to training programs or computer programs that would, at best, have an indirect emissions impact that would be extremely difficult to quantify. Clearly, identifying the most likely alternatives from among this vast array of possibilities is a highly uncertain enterprise, as is estimating the impact of the alternative project(s) on emissions. Furthermore, there is ultimately no way of reducing the uncertainty--we can never know what would have happened in the absence of the project. The uncertainty surrounding baseline estimates cannot be eliminated, and therefore some consideration should be given to accommodating this uncertainty in any future international agreements.

If market mechanisms are utilized to only a limited extent, or if market mechanism estimation errors tend to cancel out at the global level, then the differences between expected and actual emission/concentration levels should be small; in such a case, adjustment of emission reduction targets may not be necessary. Note, also, that the differences may prove to be negative rather than positive. We cannot know, in advance, whether baseline estimation errors will lead to an underestimation or an overestimation of total market-based project emission reductions. If reductions are significantly underestimated, then the estimated value of e – the factor used to adjust global emission targets to accommodate expected errors – may be negative, and the mandatory emission targets can be relaxed (thus reducing the costs of meeting emissions goal x). If on the other hand, market-based project emission reductions are overestimated, then the value of e may be positive; in this case the mandatory emission reduction goals would be tightened by an amount equal to e.

A more sophisticated alternative would be to specify e as a variable rather than a constant in future international agreements. Under this approach, future emission reduction targets would be automatically adjusted – perhaps once every 10 or 20 years – based on a pre-specified formula for computing the error adjustment factor e. Yet a third alternative would be to re-negotiate the emission reduction targets to reflect current estimates of e on a periodic basis. However, this approach has a significant disadvantage in that it would tend to politicize what should be a technical issue.

Irrespective of the particular approach chosen, it seems that some means of accommodating systemic errors may be worth considering. *In the view of the authors, the single most-important lesson to be gleaned from this project analysis exercise is the tremendous difficulty of establishing a reliable, accurate project baseline.* Numerous errors *will* occur at the project level; these errors can be reduced, but they cannot be eliminated. Furthermore, these project-level errors may well lead to significant, systemic biases in emission reduction estimates at the global level. Depending on the magnitude and direction of these biases, they have the potential for either subverting the emission reduction goals of future international agreements, or driving up the costs of market-based emission reduction efforts to the point where these efforts can no longer serve as effective cost reduction mechanisms. The high error potential inherent in the market mechanisms needs to be recognized, and perhaps accommodated, within the framework of future international agreements.

8. CONCLUSIONS AND RECOMMENDATIONS FOR FURTHER WORK

8.1 Report Summary and Conclusions

This report has presented a detailed analysis of two alternative approaches to estimating project emission baselines under a market-based mechanism environment: the project-specific approach and the modified technology matrix approach. The analysis took the form of a test or case study application of the two approaches to three example market-based projects. The project-specific approach was applied to a heat rate improvement project at coal-fired power plants in India, while the modified technology matrix was applied to an IGCC project in China and a fuel cell project in Argentina. In all three cases, the project analysis were performed from the subjective viewpoint of the project or technology matrix developers. Thus, an attempt was made to develop as persuasive an argument as possible supporting the additionality of the three projects. In addition, we sought to be as rigorous as possible in our attempts to develop accurate, reliable project baselines and benchmarks.

Finally, we subjected the three project analysis to an objective critique. Despite the rigor of the analysis, we found that the emission baselines and benchmarks established for the projects are highly uncertain. Numerous potential error sources were identified, including some of a systematic nature that could significantly bias emission reduction estimates at a global level. Some options were offered for dealing with these potential errors.

The main conclusions of this report can be summarized as follows:

- *Emission baseline estimation is a very difficult and highly uncertain process.* This, in the authors' view, is the most important conclusion to be drawn from the three project analysis. These analysis illustrate that, regardless of the approach used, there are numerous potential sources of error. Furthermore, some of these error sources are likely to cause biases in emission reduction estimates at the global level. Of particular importance, the existence of supply-demand imbalances in developing countries is a major complicating factor in project analysis, and a particularly troublesome source of potential errors.

- *Additionality classification errors are of fundamentally greater concern than baseline estimation errors.* This conclusion follows from three main considerations. First, because of asymmetry in the outcomes following upon classification errors, such errors, even if randomly distributed, will lead to the systematic overestimation of emission reductions. Second, because they are more viable than additional projects, the misclassification of non-additional projects as additional will result in preferential investment in these projects at the expense of additional projects. Third, additionality classification errors will always lead to large errors in emission reduction estimates, equal to 100 percent of the estimated project reductions. Rigorous additionality testing may thus provide the best means of guarding against large systematic biases in emission reduction estimates at the global level.

143

- *To the extent that cost-effective error reduction techniques can be applied, they can be utilized to reduce the potential for systematic errors in the estimation of emission baselines.* However, because it will prove unduly expensive to attempt to eliminate all such errors, and because, in any event, some systematic errors would almost certainly remain even given the most diligent attempts at error elimination, another more cost-effective option may be to explicitly accommodate the likelihood of market mechanism errors in future international agreements. This accommodation may take the form of constant or variable adjustment factors, to be applied to the emission reduction targets specified in future agreements.

- *Since the benchmarking approach to baseline estimation has little direct relevance to the issue of additionality, its application would probably lead to the mis-classification of large numbers of non-additional projects as additional (and vice versa).* Because non-additional projects, by definition, tend to be more viable than additional projects, project developers will exploit the benchmarking approach (knowingly or unknowingly) by preferentially investing in the misclassified non-additional projects, at the expense of truly additional projects. As a result, the number of emission reduction credits awarded will exceed the reductions actually achieved, potentially undermining emission reduction goals.

- *By combining the technology-based test for additionality with the benchmarking approach, the modified technology matrix represents a potential alternative to the benchmarking approach.* The modified technology matrix is a modified version of benchmarking, combining a technology-based test for additionality with the benchmark approach to baseline development. The modified technology matrix is applicable to all projects utilizing advanced, non-commercial technologies.

- *Because the modified technology matrix is limited in its scope of application to advanced technologies, it should not be used as an exclusive baseline methodology, but should be used in combination with another approach.* For example, the project-specific approach could be applied to projects utilizing conventional commercial technologies not covered by the technology matrix.

BIBLIOGRAPHY

Blackman, Allen and Xun Wu. "Foreign Direct Investment in China's Power Sector: Trends, Benefits and Barriers." Resources for the Future. Washington D.C. September 1998.

Center for Clean Air Policy. "JI Braintrust Group: Minutes of the February 18-19 1998 Meeting."

Center for Clean Air Policy. "JI Braintrust Group: Minutes of the May 4-5 1998 Meeting."

Chomitz, Kenneth M. "Baselines for Greenhouse Gas Reductions: Problems, Precedents, Solution." Draft Paper. World Bank, Washington D.C., July 1998.

Coal Industry Advisory Board (CIAB), International Energy Agency (IEA). "Coal in the Energy Supply of China." Report of the CIAB Asia Committee. Paris, France, 1999.

Ellis, Jane. "Experience with Emission Baselines Under the AIJ Pilot Phase." OECD Information Paper. Paris, April, 1999.

Fuel Cells 2000. "Fuel Cells 2000: Frequently Asked Questions about Fuel Cells." The Online Fuel Cells Information Center, http;//www.fuelcells.org/fuel/fcfaqs.html

Government of India, Ninth Five-Year Plan.

Graham, Kyle and Rajini Ramakrishnan. "Sustainable Development, Emissions Reduction Initiatives and the Clean Development Mechanism." Prepared by the Yale Environmental Law Clinic for the Center for Sustainable Development in the Americas, April 1999.

Hargrave, Tim, Ned Helme and Ingo Puhl. "Options for Simplifying Baseline Setting for Joint Implementation and Clean Development Mechanism Projects." Center for Clean Air Policy, Washington D.C. November, 1998

Lazarus, Michael, Sivan Kartha, Michael Ruth, Steve Bernow, and Carolyn Dunmire. "Evaluation of Benchmarking as an Approach for Establishing Clean Development Mechanism Baselines" A report to USEPA prepared by Tellus Institute, Stockholm Environment Institute, and Stratus Consulting, Inc. October 1999.

Mollot, Darren J. USDOE. "Clean Coal Technologies & Beyond: The Status and Prospects of Coal Power" Presented at the Johns Hopkins University, School of Advanced International Studies (SAIS). April 28, 1999.

Moore, Taylor. "Market Potential High for Fuel Cells" *EPRI Journal*, May/June 1997.

Nautilus Institute for Security and Sustainable Development. "IGCC in China". A background paper for the ESENA Workshop on Innovative Financing for Clean Coal in China: A GEF Technology Risk Guarantee? Berkeley, California. February 27-28, 1999.

Puhl, Ingo. "Status of Research on Project Baselines Under the UNFCCC and the Kyoto Protocol." OECD and IEA Information Paper. Paris, October 1998.

Tata Energy Research Institute, "Indian Power Sector: Change of Gear,"
 http://www.teriin,org/energy/power.htm.

U.S. Department of Commerce, International Trade Administration. "Air Pollution Control Measuring Equipment: Argentina." Market Research Reports: Industry Sector Analysis. May, 1997.

U.S. Department of State. "Submission of the United States on the Review of the Activities Implemented Jointly (AIJ) Pilot Phase." February 12, 1999.

U.S. Energy Information Administration. Argentina Country Analysis Brief. http://www.eia.doe.gov/emeu/cabs/argentina.html

U.S. Energy Information Administration (EIA), China Country Analysis Brief. June 1999. http://www.eia.doe.gov/emeu/cabs/china.html

U.S. Initiative on Joint Implementation. "Resource Document on Project and Proposal Development under the U.S. Initiative on Joint Implementation (USIJI)." Version 1.1, Washington D.C., June 1997.

White, David. "Clean Coal Technology; Buggenum; A step towards commercialization of IGCC" *Modern Power System*. June 30, 1998.

Zhesheng, Jiang. "IGCC Demonstrated Power Plant in China". Presented at the ESENA Workshop on Innovative Financing for Clean Coal in China: A GEF Technology Risk Guarantee? Berkeley, California. February 27-28, 1999.

GLOSSARY

Additionality: Refers to the issue of whether a greenhouse gas abatement or sequestration project will produce emission benefits in addition to those that would have occurred without the project.

Benchmark Baseline Approach: Baseline estimation approach in which a set of stipulated baseline emission rates are provided for different countries, sectors, or sub-sectors.

Capital stock: Property, plant and equipment used in the production, processing and distribution of energy resources.

Carbon dioxide: A colorless, odorless, non-poisonous gas that is a normal part of the ambient air. Carbon dioxide is a product of fossil-fuel combustion. Although CO_2 does not directly impair human health, it is a greenhouse gas that traps the earth's heat and contributes to the potential for global warming.

Climate: The average course or condition of the weather over a period of years as exhibited by temperature, humidity, wind velocity, and precipitation.

Counterfactual: The scenario that is expected to occur ("business-as-usual scenario") without the implementation of a clean development project.

Credits: Verified units of greenhouse gas reductions from abatement or sequestration projects to occur via market- or project-based activities.

Economic Feasibility Approach: An approach to determining additionality based on a project's economics (net present value, internal rate of return, etc.)

Electrical generating capacity: The full-load continuous power rating of electrical generating facilities, generators, prime movers, or other electric equipment (individually or collectively).

Emission Baselines: A standard from which a measure of emission reductions or carbon sequestration is established.

Emissions: Anthropogenic (human-caused) releases of greenhouse gases to the atmosphere (e.g., the release of carbon dioxide during fuel combustion).

Emissions coefficient/factor: A unique value for scaling emissions to activity data in terms of a standard rate of emissions per unit of activity (e.g., pounds of carbon dioxide emitted per barrel of fossil fuel consumed).

Flue gas desulfurization: Equipment used to remove sulfur oxides from the combustion gases of a boiler plant before discharge to the atmosphere. Also referred to as scrubbers. Chemicals such as lime are used as scrubbing media.

Fluidized-bed combustion: A method of burning particulate fuel, such as coal, in which the amount of air required for combustion far exceeds that found in conventional burners. The fuel particles are continually fed into a bed of mineral ash in the proportions of 1 part fuel to 200 parts ash, while a flow of air passes up through the bed, causing it to act like a turbulent fluid.

Fossil fuel: Any naturally occurring organic fuel formed in the Earth's crust, such as petroleum, coal, or natural gas.

Fuel cycle: The entire set of sequential processes or stages involved in the utilization of fuel, including extraction, transformation, transportation, and combustion. Emissions generally occur at each stage of the fuel cycle.

Fugitive emissions: Unintended leaks of gas from the processing, transmission, and/or transportation of fossil fuels.

Greenhouse effect: A popular term used to describe the roles of water vapor, carbon dioxide, and other gases in keeping the Earth's surface warmer than it would otherwise be. These radioactively active gases are relatively transparent to incoming shortwave radiation, but are relatively opaque to outgoing longwave radiation. The latter radiation, which would otherwise escape to space, is trapped by greenhouse gases within the lower levels of the atmosphere. The subsequent re-radiation of some of the energy back to the Earth maintains higher surface temperatures than would occur if the gases were absent. There is concern that increasing concentrations of greenhouse gases, including carbon dioxide, methane, and chlorofluorocarbons, may enhance the greenhouse effect and cause global warming.

Greenhouse gases: Those gases, such as water vapor, carbon dioxide, tropospheric ozone, nitrous oxide, and methane, that are transparent to solar radiation but opaque to longwave radiation, thus preventing longwave radiation energy from leaving the atmosphere. The net effect is a trapping of absorbed radiation and a tendency to warm the planet's surface.

Level of Error: Refers to the occurrence of error in CERs awarded, caused by incorrectly estimated emissions scenarios and/or the mis-classification of additional and non-additional activities. Low levels of error are particularly important because even randomly distributed classification errors lead to biased market-based project emission reduction estimates. Moreover, additionality classification errors always lead to emission reduction estimation errors equivalent to 100 percent of estimated project reductions.

Intergovernmental Panel on Climate Change (IPCC): A panel established jointly in 1988 by the World Meteorological Organization and the United Nations Environment Program to assess the scientific information relating to climate change and to formulate realistic response strategies.

Joint Implementation (JI): Or activities implemented jointly was introduced under the UN Framework Convention on Climate Change where industrialized countries meet their greenhouse gas emissions by receiving credits for investing in emissions reductions in developing countries.

Market- or Project- Based Mechanism or Activity:: A project between a developed country and a developing county that provides the developing county with the financing and technology for sustainable development and assists the developed country in achieving compliance with its emission reduction commitments.

Modified Technology Matrix Baseline Approach: A baseline estimation approach in which a set of technologies is pre-qualified as additional based on a consideration of their economics and current market penetration; stipulated benchmarks are then provided for each pre-qualifying technology as the basis estimating the baselines.

Moral Hazard: A concept referring to the notion that some countries may keep in place inefficient and carbon-intensive regulatory energy policies in order to increase opportunities for market-mechanism investment.

Non-Economic Barriers: An approach to determining additionality based on non-economic hindrances to project implementation such as lack of knowledge of project-related technologies.

Project-Specific Baseline Approach: A baseline estimation approach involving the tailoring of a different procedure to each individual project, based on a detailed analysis of the project's defining characteristics.

148

Renewable energy: Energy obtained from sources that are essentially inexhaustible (unlike, for example, the fossil fuels, of which there is a finite supply). Renewable sources of energy include wood, waste, geothermal, wind, photovoltaic, and solar thermal energy.

Sample: A set of measurements or outcomes selected from a given population.

Secondary Effects: Additional often unintended impacts of a carbon offset project. Also referred to as "leakage."

Sustainable Development: A broad concept referring to a country's need to balance the satisfaction of near-term interests with the protection of the interests of future generations.

Transaction Costs: The transaction costs of developing a project include cost of proposal preparation, responding to technical questions raised during project evaluation, travel costs, etc. Because smaller-sized projects are particularly affected by high transaction costs, cost-effective project certification procedures are needed to ensure maximum participation in market-based mechanisms.

Transparency: To ensure replicability and independent verification of emission reductions, transparent measures for defining and evaluating baselines are necessary. Thus, common and consistent baseline methodologies are needed, including explicitly defined information and data assumptions.

Uncertainty: A measure used to quantify the plausible maximum and minimum values for emissions from any source, given the biases inherent in the methods used to calculate a point estimate and known sources of error.

www.ingramcontent.com/pod-product-compliance
Lightning Source LLC
Chambersburg PA
CBHW081123170526
45165CB00008B/2523